HANDBOOK OF CAPTURE–RECAPTURE ANALYSIS

Ear-tagged female polar bear with 2 cubs on the sea ice 60 miles northeast of Prudhoe Bay, Alaska, 1985. (Photo by Steven C. Amstrup)

HANDBOOK OF CAPTURE–RECAPTURE ANALYSIS

Edited by

Steven C. Amstrup, Trent L. McDonald,
and Bryan F. J. Manly

PRINCETON UNIVERSITY PRESS PRINCETON AND OXFORD

Published by Princeton University Press, 41 William Street, Princeton, New Jersey 08540

In the United Kingdom: Princeton University Press, 3 Market Place, Woodstock, Oxfordshire OX20 1SY

Library of Congress Cataloging-in-Publication Data

Handbook of capture–recapture analysis / edited by Steven C. Amstrup, Trent L. McDonald, and Bryan F. J. Manly.
    p. cm.
Includes bibliographical references and index.
ISBN-13: 978-0-691-08967-6 (cl : alk. paper)
ISBN-13: 978-0-691-08968-3 (pb : alk. paper)
ISBN-10: 0-691-08967-1 (cl : alk. paper)
ISBN-10: 0-691-08968-X (pb : alk. paper)
    1. Animal populations—Mathematical models. I. Amstrup, Steven C.
II. McDonald, Trent L., 1965– III. Manly, Bryan F. J., 1944–

QL752.H36 2005
591.7'88'015118—dc22   2004065440

British Library Cataloging-in-Publication Data is available

Printed on acid-free paper. ∞

pup.princeton.edu

Printed in the United States of America

10 9 8 7 6 5 4 3 2

Contents

List of Illustrations

List of Tables

Preface

THE IDEA to produce this book was conceived at the SEEM3 Conference held in Dunedin, New Zealand, in 1999. This conference on capture–recapture methods, as well as papers appearing in professional journals and our personal experiences, convinced us that there is a wide gap between statistician and biologist in the understanding of modern capture–recapture analyses. In recent years, statisticians and mathematicians have made great strides improving our ability to make use of data collected when animals are repeatedly captured or observed. However, whereas this field of endeavor is as old as biology itself, we sensed that the appreciation many practicing biologists have for modern methods lags significantly behind the recent mathematical and statistical developments. We have assembled the *Handbook of Capture–Recapture Analysis* to guide biologists toward a greater appreciation of capture–recapture methods in theory and practice.

This handbook is organized for ease of learning and understanding. Critical introductory material and less complex methods are presented early in the book. As the chapters increase in number (through to chapter 8) so does the complexity of the material covered. Therefore, we suggest that you carefully study chapter 1, before reading anything else. After grasping the material therein, carefully read chapters 2, 3, and 4 before exploring the subsequent chapters. Once you have a solid grasp of the concepts presented in those early chapters, it will be easier to find and understand descriptions of the more complex methods presented in later chapters that may best suit your particular needs. Also, before launching into methods described here, we encourage you to examine the practical examples presented in chapter 9. These will help anchor your understanding of the material that you read in the earlier chapters. Finally, we encourage you to study the summaries in chapter 10. We feel this approach will allow you to most efficiently learn and apply the material in this volume to your particular research or management problems. As auxiliary information, a list of currently available software for capture–recapture analysis is included in an appendix.

We would like to thank Nadine Wilson, Karyn Sernka, and Kimberly Bay for their valuable help in formatting the manuscript and for compiling the references. We would also like to thank the authors of the individual chapters who have persevered with us through this process.

Western EcoSystems Technology, Inc. and the U. S. Geological Ser-
vices Alaska Science Center provided funding, in approximately equal
proportions, for the project.

Steven Amstrup, Trent McDonald, and Bryan Manly
June 2005

HANDBOOK OF CAPTURE–RECAPTURE ANALYSIS

One

Introduction to the Handbook

BRYAN F. J. MANLY, TRENT L. McDONALD,
AND STEVE C. AMSTRUP

1.1 Introduction

In September of 1802, Pierre Simon Laplace (1749–1827) used a capture–recapture type of approach to estimate the size of the human population of France (Cochran 1978; Stigler 1986). At that time, live births were recorded for all of France on an annual basis. In the year prior to September 1802, Laplace estimated the number of such births to be approximately $X = 1,000,000$. These newly born individuals constituted a marked population. Laplace then obtained census and live birth data from several communities "with zealous and intelligent mayors" across all of France. Recognizing some variation in annual birth rates, Laplace summed the number of births reported in these sample communities for the three years leading up to the time of his estimate, and divided by three to determine that there were $x = 71,866$ births per year (marked individuals) in those communities. The ratio of these marked individuals to the total number of individuals in the sampled communities, $y = 2,037,615$ was then the estimate

$$p = \frac{71,866}{2,037,615} = 0.0353$$

of the proportion of the total population of France that was newly born. On this basis, the one million marked individuals in the whole of France is related to the total population N as

$$Np \approx 1,000,000$$

so that

$$N \approx \frac{1,000,000}{0.0353} = 28,328,612$$

Figure 1.1. Jeff Mason fires a shoulder-held cannon net used to capture Bristle-thighed Curlews (*Numenius tahitiensis*) on the Seward Peninsula, spring 1989. (Photo by Robert Gill)

This estimation procedure is equivalent to the Lincoln-Peterson capture–recapture estimator described in chapter 2.

Although Laplace is commonly thought of as the first to use the capture–recapture idea, he was preceded by almost 200 years by John Graunt in his attempts to use similar methods to estimate the effect of plague and the size of populations in England in the early 1600s (Hald 1990). The theories and applications of capture–recapture have moved far beyond the concepts of John Graunt and Pierre Laplace in the ensuing centuries. Current methods do, however, share the basic concept, of ratios between known and unknown values, that guided those pioneers.

Our purpose in this book is to provide a guide for analyzing capture–recapture data that can lead the naive reader through basic methods, similar to those used by the earliest of workers, to an understanding of modern state of the art methods. This handbook is intended primarily for biologists who are using or could use capture–recapture to study wildlife populations. To the extent practicable, therefore, we have kept mathematical details to a minimum. We also have, beginning with this first chapter, attempted to explain some of the mathematical details that are necessary for a complete conceptual understanding of the methodologies described. Also, authors of each chapter have been encouraged to provide all the references that are necessary to enable readers to obtain more details about the derivations of the methods that are discussed. Therefore, this book also will be a useful introduction to this subject for statistics students, and a comprehensive summary of methodologies for practicing biometricians and statisticians.

The book is composed of three sections. Section 1 is this chapter, which is intended to set the scene for the remainder of the book, to cover some general methods that are used many times in later chapters, and to establish a common notation for all chapters. Section 2 consists of seven chapters covering the theory for the main areas of mark–recapture methods. These chapters contain some examples to illustrate the analytical techniques presented. Section 3 consists of two chapters in which we explicitly describe some examples of data sets analyzed by the methods described in chapters 2 to 8. When useful throughout the book, we discuss computing considerations, and comment on the utility of the different methods.

1.2 Overview of Chapters 2 to 8

Chapters 2 to 8 cover the main methods available for the analysis of capture–recapture models. For those who are unfamiliar with these methods the following overviews of the chapters should be useful for clarifying the relationships between them. Figure 1.2 contains a flowchart of the capture–recapture methods described in this section of the book. This flowchart may help to clarify the relationship between analyses, and will indicate the chapter (or section) containing methods appropriate for a particular data set.

Closed-population Models

A closed population is one in which the total number of individuals is not changing through births, deaths, immigration, or emigration. The first applications of capture–recapture methods were with populations that were assumed to be closed for the period of estimation. It is therefore appropriate that the first chapter in section 2 of this book should describe closed-population models. In practice, most real populations are not closed. Sometimes, however, the changes over the time period of interest are small enough that the assumption of closure is a reasonable approximation, and the effects of violating that assumption are minimal. For this reason, the analysis of capture–recapture data from closed populations continues to be a topic of interest to biologists and managers.

In chapter 2, Anne Chao and Richard Huggins begin by discussing some of the early applications of the capture–recapture method with one sample to mark some of the individuals in a population, and a second sample to

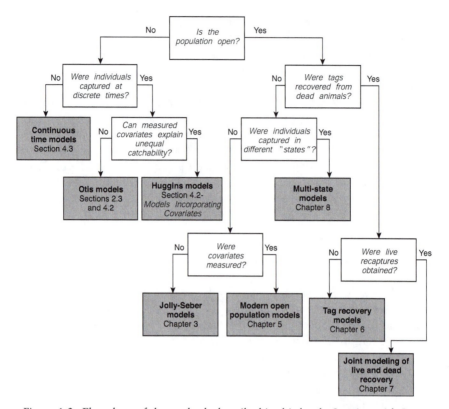

Figure 1.2. Flowchart of the methods described in this book. Starting with *Is the population open?*, unshaded boxes present Yes/No questions about the characteristics of the capture–recapture study and data. The paths induced by answers to these questions terminate at shaded boxes, which give the applicable models and this volume's chapter or section reference.

see how many marked animals are recaptured. The data obtained from the two samples can be used to estimate the population size.

A natural extension of the two-sample method, which can be traced back to Schnabel (1938), involves taking three or more samples from a population, with individuals being marked when they are first caught. The analysis of data resulting from such repeated samples, all within a time period during which the population is considered closed, is also considered in chapter 2. The goal still is estimation of the population size, but there are many more models that can be applied in terms of modeling the data. Chao and Huggins therefore conclude chapter 2 by noting the need for more general models.

The discussion is continued by Chao and Huggins in chapter 4. There they consider how the probability of capture can be allowed to vary with time, the capture history of an animal, and different animals, through the Otis et al. (1978) series of models. Other topics that are covered by Chao and Huggins in chapter 4 are the incorporation of covariates that may account for variation in capture probabilities related to different types of individuals (e.g., different ages or different sexes) or different sample times (e.g., the sampling effort or a measure of weather conditions), and a range of new approaches that have been proposed for obtaining population size estimates.

Basic Open-population Models

An open population is one that is (or could be) changing during the course of a study, because of any combination of births, deaths, immigration, or emigration. Because most natural wildlife populations are affected in this way, the interest in using capture–recapture data with open populations goes back to the first half of the 20th century when ecologists such as Jackson (1939) were sampling populations that were open, and developing methods for estimating the changing population sizes, the survival rates, and the number of individuals entering the populations between sample times.

A major achievement was the introduction of maximum likelihood estimation for the analysis of open-population capture–recapture data by Cormack (1964), Jolly (1965), and Seber (1965). This led to the development of what are now called the Cormack-Jolly-Seber (CJS) and the Jolly-Seber (JS) models. The CJS model is based solely on recaptures of marked animals and provides estimates of survival and capture probabilities only. The JS model incorporates ratios of marked to unmarked animals and thereby provides estimates of population sizes as well as survival and capture probabilities. The fundamental difference between the two is that the JS model incorporates the assumption that all animals are randomly sampled from the population and that captures of marked and unmarked animals are equally probable. The CJS model, on the other hand, does not make those assumptions and examines only the recapture histories of animals previously marked.

The CJS and JS models are the main topics of chapter 3 by Kenneth H. Pollock and Russell Alpizar-Jara. For the JS model, equations are provided for estimates of population sizes at sample times, survival rates between sample times, and numbers entering between sample times. In addition, there is a discussion of versions of this model that are restricted in various ways (e.g., assuming constant survival probabilities or constant capture probabilities) or generalized (e.g., allowing parameters to depend on the age of animals). The CJS model, which utilizes only information on the re-captures of marked animals, is then discussed. As noted above, this model has the advantage of not requiring unmarked animals to be randomly sampled from the population, but the disadvantage that this allows only survival and capture probabilities to be estimated. Population sizes, which were the original interest with capture–recapture methods, cannot be di-rectly estimated without the random sampling, which allows extrapolation from the marked to the unmarked animals in the population.

Recent Developments with Open-population Models

Since the derivation of the original CJS and JS models there have been many further developments for modeling open populations, which are covered by James D. Nichols in chapter 5. These developments are pri-marily due to the increasing availability of powerful computers, which make more flexible, but also much more complicated, modeling procedures possible. Parameter values can be restricted in various ways or allowed to depend on covariates related either to the individuals sampled or to the sample time.

The flexible modeling makes it possible to consider very large numbers of possible models for a set of capture–recapture data, particularly if the animals and sample times have values of covariates associated with them. The larger number of possible models that can be considered with modern computerized approaches elevates the importance of objective model se-lection procedures that test how well each model fits the data. It always has been necessary to assess whether models were apt, how well they fit the data, and which of the models should be considered for final selec-tion. Our greater ability now to build a variety of models is accompanied by a greater responsibility among researchers and managers to perform the comparisons necessary so that the best and most appropriate models are chosen.

The methodological developments in chapter 5 were motivated prima-rily by biological questions and the need to make earlier models more bi-ologically relevant. This underlying desire to generalize and extend the CJS model resulted in several new models. These methods, covered in chapter 5, include reverse-time modeling, which allows population growth rates to be estimated; the estimation of population sizes on the assumption that unmarked animals are randomly sampled; models that include both survival and recruitment probabilities; and the robust design in which in-tense sampling is done during several short windows of time (to meet the assumption of closure) that are separated by longer intervals of time during which processes of birth, death, immigration, and emigration may occur.

Population size estimates are derived from capture records during the short time periods of the robust design, and survival is estimated over the longer intervals between periods.

Tag-recovery Models

The tag-recovery models that are discussed by John M. Hoenig, Kenneth H. Pollock, and William Hearn in chapter 6 were originally developed separately from models for capture–recapture data. These models are primarily for analyzing data obtained from bird-banding and fish-tagging studies. In bird-banding studies, groups of birds are banded each year for several years and some of the bands are recovered from dead birds, while in fish-tagging studies, groups of fish are tagged and then some of them are recovered later during fishing operations. The early development of tag-recovery models was started by Seber (1962), and an important milestone was the publication of a handbook by Brownie et al. (1978) in which the methods available at that time were summarized.

The basic idea behind tag-recovery models is that for a band to be recovered during the jth year of a study, the animal concerned must survive for $j - 1$ years, die in the next year, and its band be recovered. This differs from the situation with capture–recapture data where groups of animals are tagged on a number of occasions and then some of them are recaptured later while they are still alive.

Joint Modeling of Tag-recovery and Live-recapture or Resighting Data

It is noted above that the difference between standard capture–recapture studies and tag-return studies is that the recaptures are of live animals in the first case, while tags are recovered from dead animals in the second case. In practice, however, the samples of animals collected for tagging do sometimes contain previously tagged animals, in which case the study provides both tag-return data and data of the type that comes from standard capture–recapture sampling.

If there are few recaptures of live animals, they will contribute little information and can be ignored. If there are many live recaptures, however, it is unsatisfactory to ignore the information they could contribute to analyses, leading to the need for the consideration of methods that can use all of the data. This is the subject of chapter 7 by Richard J. Barker, who considers studies in which animals can be recorded after their initial tagging (1) by live recaptures during tagging operations, (2) by live resightings at any time between tagging operations, and (3) from tags recovered from animals killed or found dead between tagging occasions. In addition to describing the early approaches to modeling these types of data, which go back to papers by Anderson and Sterling (1974) and Mardekian and McDonald (1981), Barker also considers the use of covariates, model selection, testing for goodness of fit, and the effects of tag loss.

Multistate Models

The original models of Cormack (1964), Jolly (1965), and Seber (1965) for capture–recapture data assumed that the animals in the population being considered were homogeneous in the sense that every one has the same probability of being captured when a sample was taken, and the same probability of surviving between two sample times. Later, the homogeneity assumption was relaxed, with covariates being used to describe different capture and survival probabilities among animals. However, this still does not allow for spatial separation of animals into different groups, with random movement between these groups. For example, consider an animal population in which members move among different geographic locales (e.g., feeding, breeding, or molting areas). Also consider that survival and capture probabilities differ at each locale. Covariates associated with the individual animals or sample times are insufficient to model this situation, and the movement between locations must be modeled directly.

In chapter 8, Carl J. Schwarz considers the analysis of studies of this type, where the population is stratified and animals can move among strata or states while sampling takes place. Analyses for these types of situations were first considered by Chapman and Junge (1956) for two samples from a closed population, extended for three samples from an open population by Arnason (1972, 1973), and to k samples from an open population by Schwarz et al. (1993b) and Brownie et al. (1993). These models can be used to study dispersal, migration, metapopulations, etc. Although the models were developed primarily to account for the physical movement of animals among geographic strata, the models described in chapter 8 also work where states are behavioral (e.g., breeding or nonbreeding animals each of which may be more or less available than the other) or habitat related rather than just geographic. Chapter 8 also shows how live and dead recoveries can be treated as different states, and describes how covariates that change randomly with time can be used to describe individuals in different states.

The first part of chapter 8 deals with the estimation of migration, capture, and survival probabilities for a stratified population, using a generalization of the Cormack-Jolly-Seber model. The last part considers the estimation of population size using two or several samples from a stratified closed population, and using several samples from an open population.

1.3 Maximum Likelihood with Capture–Recapture Methods

Early methods for analyzing capture–recapture and tag-recovery data relied upon ad hoc models for their justification. However, by the late 1960s the use of well-defined probability models with maximum likelihood estimation of the unknown parameters had become the standard approach. The method of maximum likelihood, which is known to produce estimates with good properties under a wide range of conditions, consists of two steps. First, there is the construction of a model that states the probability of observing the data as a function of the unknown parameters that are of interest. This is called the likelihood function. Second, the estimates of the unknown parameters are chosen to be those values that make the likelihood function as large as possible, i.e., the values that maximize the likelihood.

For the data considered in this book, three related types of likelihood functions need to be considered. The first and simplest arises by modeling the probability of observing the data from single independent animals, and then constructing the full likelihood as the product of probabilities for all animals. The second type of likelihood arises when data are grouped, which leads to use of the multinomial distribution to describe the probability of observing all the capture data. The third type of likelihood arises when data are collected from independent groups, which leads to likelihoods that are the product of multinomial distributions.

To illustrate the first type of likelihood, consider a four-sample experiment where n_1 animals are marked in the first sample, no more marking is done, and recapture data are obtained during samples 2, 3, and 4. Suppose that the probability of an animal surviving from the time of the jth sample to the time of the next sample is ϕ_j ($j = 1, 2,$ or 3), and the probability of a live animal being captured in the jth sample is p_j. Assume further that a particular animal was captured on the first capture occasion, and resighted on the third and fourth capture occasions. The history of captures and resightings for this animal can be indicated by the pattern of digits 1011, where a 1 in the jth position (counting from the left) indicates capture or resight during the jth occasion, and 0 in the jth position indicates that the animal was not seen during the jth occasion. Under these assumptions, the probability of observing this particular pattern of resightings, conditional on the original capture, is

$$P = \phi_1(1 - p_2)\phi_2 p_3 \phi_3 p_4 \qquad (1.1)$$

which is obtained by multiplying together the probabilities of surviving until the second sample (ϕ_1), not being captured in the second sample

$(1 - p_2)$, surviving from the second to third sample times (ϕ_2), getting captured in the third sample (p_3), surviving from the third to fourth sample times (ϕ_3), and finally getting captured in the fourth sample (p_4).

Probabilities of capture and survival for each animal in a series of samples can be used to describe the probabilities of their capture histories as in equation 1.1. The likelihood of observing all of the data is then the product of the probabilities, i.e.,

$$L = \prod_{j=1}^{n_1} P_j \qquad (1.2)$$

where P_j is the probability for the jth animal, assuming that the history for each animal is independent of the history for all of the other animals. Maximum likelihood estimation would involve finding the values of the survival and capture probabilities that maximize L. Note that ϕ_3 and p_4 cannot be estimated individually in this example because it is not possible to tell whether a large number of captured animals in the fourth and last sample is due to a high survival rate from the previous sample time or a high probability of capture. Therefore, only the product $\phi_3 p_4$ can estimated. Similarly, the capture probability cannot be estimated for the first occasion. In general, this sort of limitation applies at both ends of capture–recapture histories.

The second type of likelihood is for grouped data. In this case, the multinomial distribution is used to give the probability of the observed data. With this distribution there are m possible types of observation, with the ith type having a probability θ_i of occurring, where

$$\theta_1 + \theta_2 + \cdots + \theta_m = 1$$

If there is a total sample size of n, with n_i observations of type i occurring so that $n = n_1 + n_2 + \cdots + n_m$, then the multinomial distribution gives the probability of the sample outcome (the likelihood) to be

$$L = \frac{n!}{n_1! n_2! \cdots n_m!} \theta^{n_1} \theta^{n_2} \cdots \theta^{n_m} \qquad (1.3)$$

a probability statement that is justified in many elementary statistics texts.

Typically, when a multinomial likelihood function like this occurs in the following chapters then the θ parameters will themselves be functions of other parameters, which are the ones of real interest. For example, consider a three-sample capture–recapture study on a closed population of size N, with a capture probability of p_i for the ith sample. The possible capture–recapture patterns with their probabilities are then

$$P(000) = (1 - p_1)(1 - p_2)(1 - p_3) = \theta_1$$
$$P(100) = p_1(1 - p_2)(1 - p_3) = \theta_2$$
$$P(010) = (1 - p_1)p_2(1 - p_3) = \theta_3$$
$$P(110) = p_1 p_2(1 - p_3) = \theta_4$$
$$P(001) = (1 - p_1)(1 - p_2)p_3 = \theta_5$$
$$P(101) = p_1(1 - p_2)p_3 = \theta_6$$
$$P(011) = (1 - p_1)p_2 p_3 = \theta_7$$

and

$$P(111) = p_1 p_2 p_3 = \theta_8$$

If n_i observations are made of the ith capture–recapture pattern, then the likelihood function would be given by equation 1.3, with the θ values being functions of the p values, as shown above. In addition, because the number of uncaptured animals is unknown, this must be set equal to $n_1 = N - n_2 - n_3 - \cdots - n_8$ in equation 1.3. Maximum likelihood estimates of N, p_1, p_2, and p_3 would then be found by maximizing the likelihood with respect to these four parameters.

The third type of likelihood function occurs when the probability of the observed data is given by two or more multinomial probabilities like (1.3) multiplied together. This would be the case, for example, if the three-sample experiment just described was carried out with the results recorded separately for males and females. In that case there would be one multinomial likelihood for the males and another for the females. The likelihood for all the data would then be the product of these two multinomials. The parameters to be estimated would then be the number of males, the number of females, and capture probabilities that might or might not vary for males and females.

Likelihood Example 1

In this and the next example we illustrate some of the calculations involved in the maximum likelihood method. These examples are designed to provide the reader with a better understanding of what is meant by the phrase "estimates can be obtained by maximum likelihood" when it is used in later chapters. They are by no means a full treatment of the maximum likelihood method, but should be sufficient to provide readers with a clearer idea of the methodology behind many of the capture–recapture estimates mentioned later.

Once a likelihood for the observed data is specified, the second step in the maximum likelihood estimation process is to maximize the likelihood to obtain parameter estimates. To illustrate this second step consider again the four-sample capture–recapture experiment described above for the first type of likelihood. Suppose that $n_1 = 2$ animals are captured and marked in the first sample, and that one of these animals is recaptured in samples three and four, while the other animal is only recaptured in sample four. The capture histories for these two animals are then represented by 1011 and 1001.

Following similar logic to that used to derive equation 1.1, the probabilities of the individual capture histories are

$$P_1 = \phi_1(1 - p_2)\phi_2 p_3 \phi_3 p_4$$

and

$$P_2 = \phi_1(1 - p_2)\phi_2(1 - p_3)\phi_3 p_4$$

Assuming that the results for the two captured animals were independently obtained, the full likelihood of obtaining both the capture histories is

$$L = \prod_{j=1}^{2} P_j = [\phi_1(1 - p_2)\phi_2 p_3\, \phi_3 p_4][\phi_1(1 - p_2)\phi_2(1 - p_3)\phi_3 p_4]$$

Typically, the natural logarithm of L is taken at this point because L and $\ln(L)$ are maximized by the same parameter values, and the logarithmic function $\ln(L)$ is generally easier to maximize on a computer. The log-likelihood for this example is

$$\ln(L) = \ln(P_1) + \ln(P_2)$$
$$= \ln(\phi_1) + \ln(1 - p_2) + \ln(\phi_2) + \ln(p_3) + \ln(\phi_3) + \ln(p_4) + \ln(\phi_1)$$
$$+ \ln(1 - p_2) + \ln(\phi_2) + \ln(1 - p_3) + \ln(\phi_3) + \ln(p_4)$$
$$= 2[\ln(\phi_1) + \ln(1 - p_2) + \ln(\phi_2) + \ln(\phi_3) + \ln(p_4)] + \ln(p_3)$$
$$+ \ln(1 - p_3)$$

The process of "maximizing the likelihood" essentially entails repeatedly modifying the values of ϕ_i and p_i until $\ln(L)$ cannot be increased any more. To start the process, a set of initial parameters is defined. In this example, suppose that the maximization process is started with $\phi_i = 0.5$ for all i, and $p_i = 0.5$ for all i. Putting these initial values into $\ln(L)$ gives

$$\ln(L) = \ln(0.5) + \ln(0.5) + \ln(0.5) + \ln(0.5) + \ln(0.5) + \ln(0.5)$$
$$+ \ln(0.5) + \ln(0.5) + \ln(0.5) + \ln(0.5) + \ln(0.5) + \ln(0.5)$$
$$= 12 \ln(0.5)$$
$$= -8.32$$

It is then possible to get progressively closer to the final maximum by a judicious choice of the changes to make to the parameters. In particular, using the theory of calculus it is possible to determine the direction to change each of the parameters so that $\ln(L)$ will increase. However, the magnitude of the changes that will assure that the new values produce the maximum is not known. Consequently, small changes in the parameters are made until further changes will not increase $\ln(L)$.

The technique relies on the calculation of the derivatives of $\ln(L)$ with respect to the parameters, to specify which changes in the parameters will increase $\ln(L)$. These details are explained in texts on calculus, but are unnecessary here. For illustrating the calculations, all one needs to know is that the derivatives for the example being considered specify that changing the parameter estimates to $\phi_1 = 0.55$, $\phi_2 = 0.55$, $\phi_3 = 0.55$, $p_2 = 0.45$, $p_3 = 0.50$, and $p_4 = 0.55$ will increase $\ln(L)$. To check this, these values can be used to calculate $\ln(L)$, which gives

$$\ln(L) = \ln(0.55) + \ln(0.55) + \ln(0.55) + \ln(0.5) + \ln(0.55) + \ln(0.55)$$
$$+ \ln(0.55) + \ln(0.55) + \ln(0.55) + \ln(0.5) + \ln(0.55) + \ln(0.55)$$
$$= 10 \ln(0.55) + 2 \ln(0.5)$$
$$= -7.36$$

Repeating the process of calculating the gradient and changing the parameter estimates will eventually maximize $\ln(L)$. For example, the new derivatives calculated at the last parameter values specify that changing the parameter estimates to $\phi_1 = 0.59$, $\phi_2 = 0.59$, $\phi_3 = 0.59$, $p_2 = 0.41$, $p_3 = 0.50$ and $p_4 = 0.59$ will increase $\ln(L)$. The $\ln(L)$ value with these new parameter estimates is -6.66.

In this particular example, the likelihood is overparameterized because there are six parameters and only two capture histories. Overparameterization causes a number of problems, and, in particular, means that some parameters must be assigned arbitrary values to fix the other parameters. This overparameterized likelihood was used here, however, only to illustrate calculation of individual $\ln(L)$ values. In the next section, more capture histories are used and a more complicated likelihood function is illustrated.

Likelihood Example 2

In the last example, two capture histories were used to illustrate calculation of individual $\ln(L)$ values. In this section, the more complicated likelihood function of the Cormack-Jolly-Seber (CJS) model that is described in detail in chapter 3 and 5 is used to illustrate the process of maximizing the likelihood.

Consider the situation where animals in an open population are captured, marked, and released back into the population at each of eight capture occasions. To define the CJS likelihood, parameters p_j and ϕ_j from the previous section are needed, plus an additional parameter for the probability that an animal is never seen after a certain sample occasion. Recall that p_j is the probability that an animal in the population is captured or observed at sampling occasion j, and that ϕ_j is the probability that an animal in the population survives from sampling occasion j to $j + 1$. The new parameter needed for this situation will be called χ_j. It is the probability that an animal is not caught or seen after sampling occasion j.

Consider the capture history 01011000. Under the CJS model, the probability of this capture history occurring, conditional on the first capture, is

$$P = \phi_2(1 - p_3)\phi_3 p_4 \phi_4 p_5 \chi_5$$

The first part of this expression, $\phi_2(1 - p_3)\phi_3 p_4 \phi_4 p_5$, is justified as in earlier expressions of this type, so that a ϕ_j occurs for each interval between the first and last sampling occasions when the animal was captured, a p_j parameter occurs for each occasion that the animal is captured or seen, and a $(1 - p_j)$ term occurs for each occasion that the animal is not captured or seen. The second part of P represents the probability that the animal was not seen after occasion 5, and is represented by the parameter χ_5.

The χ_j parameters are, in fact, functions of the ϕ_j and p_j parameters. To see this, consider the eight-sample capture–recapture study. By definition,

$\chi_8 = 1$ because there is no possibility of capturing an animal after sample eight. If an animal was last seen in the seventh sample, then the probability of not seeing it in the eighth sample is the probability that the animal died, plus the probability that it lived to the time of sample eight but eluded capture, i.e.,

$$\chi_7 = (1 - \phi_7) + \phi_7(1 - p_8)$$

If an animal was last seen in the sixth sample, then the probability of not seeing the animal in the seventh or eighth sample is the probability that the animal died between the times of the sixth and seventh samples, plus the probability that it survived to the time of sample seven but eluded capture and then subsequently either died between the times of the seventh and eighth samples or eluded capture in the eighth sample, so that

$$\chi_6 = (1 - \phi_6) + \phi_6(1 - p_7)[(1 - \phi_7) + \phi_7(1 - p_8)]$$
$$= (1 - \phi_6) + \phi_6(1 - p_7)\chi_7$$

In fact, χ_j for any j can be calculated in the same way using the general recursive formula

$$\chi_j = 1 - \phi_j + \phi_i(1 - p_j)\chi_{j+1}$$

Now suppose that eight samples are taken and the capture histories

10100000	11000000
10001000	00010100
10100000	01000000
11000000	00101000
10000000	00000110
11000000	00001100
11100000	00010001

are obtained for 14 animals. Suppose further that it is assumed that the probability of survival was constant during the study and that the probability of capture was constant for all samples. If P_i is the probability of capture history i occurring under the CJS model, then the full log likelihood for this set of capture histories is the sum of $\ln(P_i)$ for $i = 1$ to 14, i.e., $\ln(L) = \Sigma \ln(P_i)$. If the constant probability of survival parameter is 0.6 and the constant probability of capture parameter is

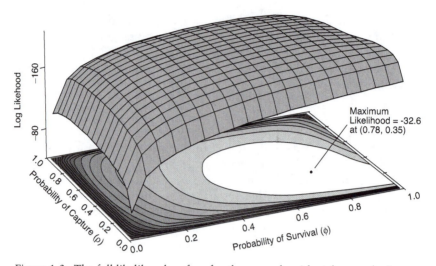

Figure 1.3. The full likelihood surface for the example with eight sample times and information on the captures and recaptures of 14 animals.

0.2, then the log likelihood for this set of data could be calculated by setting $\phi_1 = \phi_2 = \cdots = \phi_8 = 0.6$ and $p_2 = p_3 = \cdots = p_8 = 0.2$ in the CJS expression for P_j, taking the logarithms, and summing. If these calculations are carried out then it is found that $\ln(L) = -37.94$. If the probability of survival were changed to 0.65 and probability of capture changed to 0.25, then $\ln(L) = -35.18$. According to the theory of maximum likelihood, the parameters $\phi = 0.65$ and $p = 0.25$ have a higher likelihood and are therefore better than the parameters $\phi = 0.60$ and $p = 0.20$.

A computer can be programmed to repeatedly improve estimates of the parameters until $\ln(L)$ reaches a point where it cannot be increased further. For example, the SOLVER routine in Microsoft Excel can be used for this purpose providing that the likelihood function is not too complicated. With the example set of data, $\ln(L)$ will eventually reach a maximum of -32.60 when the survival parameter is $\phi = 0.78$ and the capture probability parameter is $p = 0.35$. Because this example involves only two parameters, the entire likelihood surface is easy to plot and visualize, as shown in figure 1.3.

It is also possible to estimate the standard errors of parameter estimates from the likelihood function. The mathematical details justifying these estimates involve the second derivatives of $\ln(L)$, and will not be covered here. It suffices to say that the curvature of the likelihood provides some indication about the variance of the maximum likelihood estimate. For example, figure 1.3 shows that the likelihood is relatively flat for ϕ between 0.4 and 0.9, and p between 0.15 and 0.6. It can therefore be argued that any set of parameters in this flat region of the likelihood is reasonable. In general, if the likelihood has a flat region that is large, then the maximum likelihood estimates have large variances, but if the likelihood does not have a flat region, or the flat region is small, then the maximum likelihood estimates have small variances.

1.4 Model Selection Procedures

With the flexible modeling procedures that have become possible in recent years there has been a considerable increase in the number of models that can be considered for data sets with many sampling occasions. For example, with an open population it is often the case that capture and survival probabilities can be allowed to vary with time, the sex of the animal, weather conditions, etc. The problem is then to choose a model that gives an adequate representation of the data without having more parameters than are really needed.

There are two results that may be particularly useful in this respect. First, suppose that two alternative models are being considered for a set of data. Model 1 has I estimated parameters, and a log-likelihood function of $\ln(L_1)$ when it is evaluated with the maximum likelihood estimates of the parameters. Model 2 is a generalization of model 1, with the I estimated parameters of model 1 and another J estimated parameters as well, and it has a log-likelihood function of $\ln(L_2)$ when evaluated with the maximum likelihood estimates of the parameters. Because model 2 is more general (e.g., more complex) than model 1, it will be the case that $\ln(L_2)$ is less than or equal to $\ln(L_1)$. However, if in fact the extra J parameters in model 2 are not needed and have true values of zero, then the reduction in the log likelihood in moving from model 1 to model 2,

$$D = 2[\ln(L_1) - \ln(L_2)] \tag{1.4}$$

will approximate a random value from a chi-squared distribution with J degrees of freedom. Consequently, if D, the difference or deviance, is significantly large in comparison with values from the chi-squared distribution, then this suggests that the more general model 2 is needed to properly describe the data. If, on the other hand, D is not significantly large, then this suggests that the simpler model 1 is appropriate.

A limitation with the test just described is that it applies only to nested models, that is, where one model is a special case or subset of another model. This has led to the adoption of alternative approaches to model selection that are based on Akaike's information criterion (AIC) (Akaike 1973; Burnham and Anderson 1998).

In its simplest form, AIC model selection involves defining a set of models that are candidates for being chosen as the most suitable for the data. Each model is then fitted to the data and its corresponding value for

$$AIC = -2\ln(L) + 2P \tag{1.5}$$

is obtained, where L is the maximized likelihood for the model, and P is the number of estimated parameters. The model with the smallest value for AIC is then considered to be the "best" in terms of a compromise between the goodness of fit of the model and the number of parameters that need to be estimated. The balancing of model fit and number of parameters in the model is important in determining the precision of the estimates derived.

A further comparison between models can be based on calculating Akaike weights (Buckland et al. 1997). If there are M candidate models then the weight for model i is

$$w_i = \frac{\exp(-\Delta_i/2)}{\exp(\Delta_1/2) + \exp(\Delta_2/2) + \cdots + \exp(\Delta_M/2)} \tag{1.6}$$

where Δ_i is the difference between the AIC value for model i and the smallest AIC value for all models. The Akaike weights calculated in this way are used to measure the strength of the evidence in favor of each of the models, with a large weight indicating high evidence.

There are some variations of AIC that may be useful under certain conditions. In particular, for small samples (less than 40 observations per parameter) a corrected AIC can be used, which is

$$AIC_c = AIC + \frac{2P(P+1)}{(n-P-1)} \tag{1.7}$$

where n is the number of observations and P is the number of estimated parameters. Also, if there is evidence that the data display more variation than expected based on the probability model being used then this can be allowed for using the quasi-AIC values

$$QAIC = \frac{-2\ln(L)}{\hat{c}} + 2P \qquad (1.8)$$

where $\hat{c}$ is an estimate of the ratio of the observed amount variation in the data to the amount of variation expected from the probability model being assumed, as explained more fully in chapter 9. The method for obtaining the estimate $\hat{c}$ depends on the particular circumstances. When more variation than expected under a certain model is displayed, the data are said to be "overdispersed." Overdispersion can arise in a number of ways. The most common causes are model misspecification (lack of fit) and a lack of true independence among observations. For example, the statistical likelihood for a set of capture data may assume that capture histories follow a multinomial distribution with a particular set of probabilities. If there is more variation in the capture histories than predicted by the multinomial distribution, the probabilities assumed in the multinomial model are incorrect, implying that the covariate model is misspecified, or there may be unaccounted for dependencies among the histories. In some but not all cases, apparent overdispersion can be remedied by incorporating more or different covariates into the model. Often, however, it will not be possible to account for some amount of overdispersion in the data.

1.5 Notation

A good deal of notation is necessary for describing the models used in the remainder of this book. The variation in notation can be quite confusing, particularly if sections of the book are read in isolation. To help reduce this confusion, all authors have standardized their notations, to the maximum extent practicable. In table 1.1 we have provided a summary of most of the notation used in the volume.

TABLE 1.1
Partial list of notation used throughout the book

Symbol	Definition
i	An index for individual animals. Example: h_i denotes the capture history for the ith animal.
j	An index for capture occasions (sample times). Example: p_j is the probability of capture for the jth sample.
k	The number of capture occasions (samples)
N	A population size
R	A number of animals released
m or n	The number of animals with a certain characteristic. Examples: m_h is the number of animals with capture history h, and n_j is the number of animals captured in sample j.
h	A capture history. example: $h = 001010$.
ϕ	An apparent survival probability
p	A capture probability
γ	A seniority probability
E	A probability of emigration
χ	The probability of not being seen after a trapping occasion
ξ	The probability of not being seen before a trapping occasion
M	The number of marked animals in the population
S	A pure survival probability (not involving the probability of emigration); also, the number of strata in a multistrata model
F	A probability of not emigrating, equal to $1 - E$.
R	A reporting probability
ρ	A resighting probability
F	A tag recovery probability, equal to $r(1 - S)$ when there is no emigration
ψ	A probability of moving between strata from one sampling occasion to the next for a multistrata model

Note. In some cases, symbols not listed here may be defined to represent different things in different chapters. For example, in chapter 6 S represents a pure survival probability that does not include probability of emigration, while in chapter 8 S represents the number of strata in a multistrata model. These cases have been kept to a minimum and the meaning of each symbol is clear from the context. Symbols listed here can be subscripted or superscripted as needed. For example, N_i might denote the sample size at the time of the ith sample time. Also, a caret is often used to indicate an estimate, so that $\hat{N}$ indicates an estimate of N.

Figure 1.4. Black bear (*Ursus americanus*), near Council, Idaho, 1973, wearing numbered monel metal ear tags. (Photo by Steven C. Amstrup)

Two

Classical Closed-population Capture–Recapture Models

ANNE CHAO AND RICHARD M. HUGGINS

2.1 Introduction

This chapter reviews the classical models for closed populations (i.e., for situations where the individuals in a population remain the same while it is being studied) and the history of their development. The data structure and necessary notation are introduced in section 2.2 via a small data set on the captures of deer mice. Section 2.3 reviews classical two-occasion and multiple-occasion models, i.e., models for situations where two or more than two separate samples of animals are taken. Section 2.4 summarizes the limitations of these classical models and provides motivations for the more general models that are considered in chapter 4.

As noted in chapter 1, the idea of the two-occasion capture–recapture method can be traced to Pierre Laplace, who used it to estimate the human population size of France in 1802 (Cochran 1978), and even earlier to John Graunt who used the idea to estimate the effect of plague and the size of the population of England in the 1600s (Hald 1990). Le Cren (1965) noted that other early applications of this method to ecology included Petersen's and Dahl's work on sampling fish populations in 1896 and 1917, respectively, because they recognized that the proportion of previously marked fish captured by fishermen constituted a basis for estimating the population size of the fish. Lincoln was the first to apply the method to wildlife in 1930 when he used returned leg bands from hunters to estimate duck numbers.

Important contributors to the theory of classical closed capture–recapture methods include Schnabel, Darroch, Bailey, Moran, Chapman, and Zippin, among others. The work by Schnabel (1938) and Darroch (1958, 1959) provided the mathematical framework for the models. Detailed historical developments and applications are provided in Cormack (1968), Otis et al. (1978), White et al. (1982), Pollock (1991), Seber (1982, 1986, 1992, 2001), Schwarz and Seber (1999), and Williams et al. (2002).

Figure 2.1. Endangered Néné goose (*Nesochen sandvicensis*) wearing double leg bands. Island of Maui, Hawaii, 1998. (Photo by Steven C. Amstrup)

2.2 Structure of Capture–Recapture Experiments and Data

The raw data from closed capture–recapture experiments are the capture records of all the individuals observed in a study. These are arranged in a capture history matrix, as illustrated in table 2.1 with deer mice (*Peromyscus* sp.) data collected by V. Reid. These data are displayed in the format appropriate for CAPTURE, a widely used computer program for the analysis of closed models (Otis et al. 1978; White et al. 1982; Rexstad and Burnham 1991; section 4.4 of this book). The data arose from a live-trapping experiment that was conducted for six consecutive nights (columns) with a total of 38 mice (rows) captured over these six capture occasions. The time period for the experiment was relatively short and it was reasonable to assume that the population was closed.

In table 2.1, the capture history of each captured individual is expressed as a series of 0's (noncaptures) and 1's (captures). Thus, in the example the capture history matrix consists of 38 rows and 6 columns, with the rows representing the capture histories of each captured individual and the columns representing the captures on each occasion. The first mouse, with capture record 111111, was captured on all six nights. The second mouse,

TABLE 2.1
Individual capture history of 38 deer mice with six capture occasions

Occasion 1	Occasion 2	Occasion 3	Occasion 4	Occasion 5	Occasion 6
1	1	1	1	1	1
1	0	0	1	1	1
1	1	0	0	1	1
1	1	0	1	1	1
1	1	1	1	1	1
1	1	0	1	1	1
1	1	1	1	1	0
1	1	1	0	0	1
1	1	1	1	1	1
1	1	0	1	1	1
1	1	0	1	1	1
1	1	1	0	1	1
1	1	1	1	1	1
1	0	1	1	1	0
1	0	0	1	0	0
0	1	0	0	1	0
0	1	1	0	0	1
0	1	0	0	0	1
0	1	0	1	0	1
0	1	1	0	1	0
0	1	0	1	0	1
0	1	0	0	0	1
0	1	0	0	1	1
0	0	1	0	0	0
0	0	1	1	1	1
0	0	1	0	1	1
0	0	1	1	1	1
0	0	1	0	1	0
0	0	1	0	0	0
0	0	0	1	0	0
0	0	0	1	1	1
0	0	0	1	1	0
0	0	0	0	1	0
0	0	0	0	1	0
0	0	0	0	1	0
0	0	0	0	0	1
0	0	0	0	0	1
0	0	0	0	0	1

with capture record 100111, was captured on nights 1, 4, 5, and 6, but not on nights 2 and 3. Similar interpretations apply to other capture histories.

In larger studies with numerous capture occasions and many captured individuals, the capture-history matrix becomes very large, and with the classical models it is more convenient to represent the raw data by a tally of the frequencies of each capture history, which retains most of the information in the original capture-history matrix.

For many classical estimation procedures, the following summary statistics are sufficient for the statistical analysis:

k = the number of capture occasions;

n_j = the number of animals captured on the jth capture occasion, $j = 1, \ldots, k$;

u_j = the number of unmarked animals captured on the jth capture occasion, $j = 1, \ldots, k$;

m_j = the number of marked animals captured on the jth capture occasion, $j = 1, \ldots, k$, where $m_1 = 0$;

M_j = the number of distinct animals captured before the jth capture occasion, $j = 1, \ldots, k$, where this is the same as the number of marked animals in the population just before the jth capture occasion, and of necessity $M_1 = 0$ and M_{k+1} is defined as the total number of distinct animals captured in the experiment; and

f_j = the number of animals captured exactly j times, $j = 1, \ldots, k$.

These statistics are given in table 2.2 for the data in table 2.1. The statistic n_j denotes the column sum for the jth column (occasion) in the capture history matrix, with $(n_1, n_2, \ldots, n_6) = (15, 20, 16, 19, 25, 25)$. Out of the n_j animals, there are u_j first captures and m_j recaptures, so that $u_j + m_j = n_j$, with $(u_1, u_2, \ldots, u_6) = (15, 8, 6, 3, 3, 3)$, and $(m_1, m_2, \ldots, m_6) = (0, 12, 10, 16, 22, 22)$. The statistic M_j can also be interpreted as the cumulative number of first-captures on the first $j - 1$ occasions, thus $M_j = u_1 + u_2 + \cdots + u_{j-1}$ and $(M_1, M_2, \ldots, M_7) = (0, 15, 23, 29, 32, 35, 38)$. That is, the number of marked individuals in the population progressively increased from $M_1 = 0$ to $M_7 = 38$.

The row sum for each individual denotes the capture frequency of that animal, and $(f_1, f_2, \ldots, f_k)$ represent the frequency counts of all captured animals. As shown in table 2.2, the frequency counts for the mouse data are $(f_1, f_2, \ldots, f_6) = (9, 6, 7, 6, 6, 4)$. That is, 9 animals were captured once, 6 animals captured twice, . . . and 4 animals captured on all 6 occasions. The term f_0 is the number of animals never captured, so that $f_1 + f_2 + \cdots + f_k = M_{k+1}$ and $f_0 + f_1 + \cdots + f_k = N$. Therefore, estimating the population size N is equivalent to estimating the number of missing animals, f_0.

TABLE 2.2
Summary statistics for the deer mouse data

j	n_j	m_j	u_j	M_j	f_j
1	15	0	15	0	9
2	20	12	8	15	6
3	16	10	6	23	7
4	19	16	3	29	6
5	25	22	3	32	6
6	25	22	3	35	4
				$(M_7 = 38)$	
Total	120	82	38		38

2.3 Early Models and Estimators

Some assumptions are common to all closed-population models:

1. The population remains constant over the study period (e.g., there is no immigration or emigration), although known removals (e.g., deaths on capture) are allowed (the closure assumption).
2. Animals do not lose their marks or tags.
3. All marks or tags are correctly recorded.
4. Animals act independently.

Two-occasion Models (Petersen-Lincoln Estimator)

The two-occasion case is the origin of capture–recapture methodology. The intuitive idea is the following: Assume that on the first capture occasion, a sample of n_1 animals is captured, marked, and released. Therefore, the marked proportion of the population is

$$\frac{n_1}{N}$$

On the second capture occasion, n_2 animals are caught, m_2 of which are marked. The proportion of the marked animals in the second sample is m_2/n_2. Assuming that the marked proportion in the sample is equal to the marked proportion in the population suggests that

$$\frac{m_2}{n_2} \approx \frac{n_1}{N}$$

which yields the estimate of the population size

$$\hat{N}_P = \frac{n_1 n_2}{m_2} \tag{2.1}$$

This is the well-known Petersen-Lincoln estimator, which is also known as the Lincoln-Petersen estimator.

To illustrate this estimator, consider the first two occasions in table 2.1. From table 2.2, $n_1 = 15$, $n_2 = 20$, and $m_2 = 12$, with a total of 23 distinct mice being captured on the first two occasions. The Petersen-Lincoln estimator is then

$$\hat{N}_P = \frac{15 \times 20}{12} = 25$$

It appears that this estimate based on capture data from only the first two occasions is not sensible, since a total of 38 distinct mice were actually seen at the end of the experiment.

Various sampling models have been proposed to justify the Petersen-Lincoln estimator and obtain estimates of the associated standard error. A commonly used model assumes a multinomial distribution (section 1.3) under the assumption that on each occasion all animals have the same capture probability (equal catchability). With the multinomial model, the number of captures on each occasion is a random variable. There are three observable capture histories (10, 01, and 11) and three parameters (N, p_1, p_2), where p_1 and p_2 denote the capture probabilities for the first and second capture occasions, and the Petersen-Lincoln estimator turns out to be an approximate maximum likelihood estimator (MLE) of N (Seber 1982, p. 131).

Another approach uses the hypergeometric model, and treats the sample sizes n_1 and n_2 as fixed. This model contains only one parameter N and one random variable m_2. Also, under the assumption of equal catchability, the MLE of N is the integer part of the Petersen-Lincoln estimator. If $n_1 + n_2 \geq N$, then an unbiased estimator proposed by Chapman (1951) is

$$\hat{N}_C = \frac{(n_1 + 1)(n_2 + 1)}{m_2 + 1} - 1 \tag{2.2}$$

An intuitive explanation for why this estimator can reduce bias is that the bias of the Petersen-Lincoln estimator is mainly due to small values of m_2, especially $m_2 = 0$. This means no animals are captured twice, and the estimate of equation 2.1 is infinite. Hence, one extra animal is added into the frequency of animals caught on both occasions so that m_2 is at least one, and n_1, n_2 and the population size are consequently all increased by one. Chapman's estimator follows from the fact that $(n_1 + 1) \times (n_2 + 1)/(m_2 + 1)$ is a valid estimator for $N + 1$. When $n_1 + n_2 < N$, the bias of Chapman's estimator is approximately equal to $-N \exp[-(n_1 + 1) \times (n_2 + 1)/N]$ (Seber 1982, p. 60). There are other bias-corrected estimators. For example, Bailey (1951, 1952) adopted a binomial model and obtained a different bias-corrected estimator.

The difference between the Petersen-Lincoln and Chapman's estimator is small when the number of recaptures is not small relative to n_1 and n_2. For example, Chapman's estimate based on the first two capture occasions of table 2.1 is

$$\hat{N}_C = \frac{(n_1 + 1)(n_2 + 1)}{(m_2 + 1)} - 1 = \frac{16 \times 21}{13} - 1 = 24.85$$

compared to the Petersen-Lincoln estimate of 25.

Both estimators have approximately the same variance formula, given by

$$\text{Var}(\hat{N}_C) \approx \text{Var}(\hat{N}_P) \approx \frac{(n_1 + 1)(n_2 + 1)(n_1 - m_2)(n_2 - m_2)}{(m_2 + 1)^2(m_2 + 2)} \qquad (2.3)$$

(Seber 1982, p. 60). Hence, the variances are approximately $16 \times 21 \times 3 \times 8/(13 \times 13 \times 14) = 3.41$ and the approximate standard error is $\sqrt{3.41} = 1.85$. A traditional 95% confidence interval based on Chapman's estimator and the large-sample normal approximation can therefore be calculated as $24.85 \pm 1.96 \times 1.85 = (21, 29)$. This interval clearly cannot cover the true population size because 38 animals were actually captured in the experiment. Suppose that only the data for the first two occasions are available, a drawback for the traditional confidence interval is that the lower limit of this interval is less than the observed number of animals, which is 23 in this case. A modification is to replace the lower bound by the number of captured. That is, the resulting interval is modified to (23, 29) based on the first two occasions. More discussion on interval estimation is presented in chapter 4.

Multiple-recapture Models (The Schnabel Census)

As noted above, two capture occasions may not result in reliable estimates. Two-occasion models were extended to multiple occasions by Schnabel

(1938) and Darroch (1958). Their approaches formed the basis of the classical models. Like the Petersen-Lincoln model, these early models assume that on each sampling occasion all animals have the same probability of capture, although this probability can be allowed to vary among sampling occasions. The models have been extensively discussed in the literature because the capture probabilities may vary with environmental conditions or the sampling effort. Because the sample data taken on the k occasions are statistically independent under the assumption of equal catchability, this model is also referred to as an "independent" model in the literature.

A special case of the equal catchability model is where the capture probability p is constant over the capture occasions. This is usually referred to as model M_0, the subscript 0 referring to no variation. Then, for example, for the data in table 2.1, the probability of an animal having the capture history 111111 is p^6; the probability of an animal having the capture history 100111 is $p^4(1-p)^2$, and the probability an animal is not captured is $(1-p)^6$. As discussed in section 1.3, by considering all possible capture histories and the corresponding counts, the likelihood based on a multinomial model becomes in this case

$$L(N,p) = \frac{N!}{(N-M_{k+1})! \prod_h a_h!} p^{n.}(1-p)^{kN-n.} \tag{2.4}$$

where a_h denotes the frequency for observable capture history h and $n. = n_1 + n_2 + \cdots + n_k$ denotes the total number of captures, which is 120 for the example.

In model M_0, there are only the two parameters, N and p, and the statistics that are necessary to estimate them are M_{k+1} and $n.$. The MLE of N and p can be obtained using numerical methods that maximize equation 2.4. A large-sample variance (Darroch 1958) is obtained by substituting the estimates into the formula

$$\mathrm{Var}(\hat{N}) = \frac{N}{(1-p)^{-k} + (k-1) - k(1-p)^{-1}}$$

For the data in table 2.1, the MLE are $\hat{p} = 0.526$ and $\hat{N} = 38$ (standard error = 0.69). After model M_0 is applied it is concluded that nearly all animals were caught, with a narrow 95% confidence interval of (38, 39.4) for the true population size.

This simple model, however, does not fit the data properly. Although more rigorous statistical tests (to be discussed below) are not possible for this data set due to the sparse counts, the fitted (expected) values shown

TABLE 2.3
Fitted (expected) values for deer mouse data for two models

	Model M_0			Model M_t		
j	$E(n_j)$	$E(m_j)$	$E(u_j)$	$E(n_j)$	$E(m_j)$	$E(u_j)$
1	20	0	20.0	15.0	0	15.00
2	20	10.53	9.47	20.0[a]	7.90[b]	12.11[c]
3	20	15.51	4.49	16.0	11.41	4.59
4	20	17.87	2.13	19.0	15.85	3.15
5	20	18.99	1.01	25.0	22.93	2.08
6	20	19.52	0.48	25.0	24.29	0.71
Total	120	82.42	37.58	120	82.38	37.64

Examples of calculations:

[a] The fitted size for sample 2 is $Np_2 = 38 \times 0.526 = 20.0$, with estimates used for N and p_2. The same equation applies with model M_0 but the same p value applies for all samples.

[b] The fitted number of marked animals in sample 2 is Np_1p_2, with estimates used for N, p_1, and p_2. The same equation applies with model M_0 but with a constant p value for all samples.

[c] The fitted number of unmarked animals in sample 2 is the difference between the fitted sample size and the fitted number of marked animals. The same holds for model M_0.

in table 2.3 do help to check the model. Because the model assumes that the capture probabilities are the same for all k occasions, the fitted values for the numbers of captures on the six occasions are equal, i.e., $E(n_1) = E(n_2) = \cdots = E(n_6) = 20$. The other fitted values follow from the formulas $E(u_j) = N(1 - p)^{j-1}p$ and $E(m_j) = N[1 - (1 - p)^{j-1}]p$. With model M_0, it is expected that the number of new animals captured will decrease rapidly from 20 to 0.5, but the data did not show such a trend. The counts $(u_1, u_2, \ldots, u_6, f_0)$ follow a multinomial distribution, and thus the usual chi-squared goodness-of-fit test can be applied. Pooling u_4, u_5, u_6, and f_0, we have four observed cells $(u_1, u_2, u_3, u_4 + u_5 + u_6 + f_0) = (15, 8, 6, 9)$. Here $f_0 = 0$ because the population size estimate is 38. Note that $E(u_4 + u_5 + u_6 + f_0) = \hat{N} - E(u_1 + u_2 + u_3)$, and it follows from table 2.3 that the corresponding expected values are $(20.0, 9.47, 4.49, 4.04)$ with $\hat{N} = 38$. Then the chi-squared value is 8.08 with 1 degree of freedom. The p value is 0.004, which reveals that the model is not adequate. Here the degree of freedom for the test is equal to the number of cells $-1 - r$, where r denotes the number of parameters, which is $r = 2$ under model M_0.

If we modify the classical model to allow capture probabilities to vary with time, we refer to it as model M_t, where the subscript t refers to variation of the capture probabilities over time. Denote the capture probability for all individual animals on the jth occasion by p_j. The likelihood for the special case of $k = 3$ was presented in section 1.3. The extension to any number of occasions is similar. Unlike model M_0, where only one parameter is used to model the capture probability, if there are k capture occasions, then model M_t uses the k parameters $p_1, p_2, \ldots, p_k$ to describe the capture probabilities. For example in table 2.1, the probability for an animal having capture history 111111 is $p_1 p_2 p_3 p_4 p_5 p_6$; the probability for an animal having capture history 100111 is $p_1(1 - p_2) \times (1 - p_3)p_4 p_5 p_6 \cdots$; and the probability an animal is not captured is $(1 - p_1)(1 - p_2)(1 - p_3)(1 - p_4)(1 - p_5)(1 - p_6)$. The multinomial likelihood representing an extension of equation 2.1 turns out to be

$$L(N, p_1, \ldots, p_k) = \frac{N!}{(N - M_{k+1})! \prod_h a_h!} \prod_{j=1}^{k} [p_j^{n_j}(1 - p_j)^{N - n_j}] \qquad (2.5)$$

There are $k + 1$ parameters $(N, p_1, p_2, \ldots, p_k)$ and the statistics needed to estimate these are $(n_1, n_2, \ldots, n_k, M_{k+1})$. The iterative procedure for finding the exact MLE is described by Otis et al. (1978, p. 106) and the variance is then approximately given by

$$\text{Var}(\hat{N}) = \frac{N}{\prod_{j=1}^{k}(1 - p_j)^{-1} + (k - 1) - \sum_{j=1}^{k}(1 - p_j)^{-1}} \qquad (2.6)$$

For the data of table 2.1, the capture probabilities are respectively estimated by $\hat{p}_1 = 0.395$, $\hat{p}_2 = 0.526$, $\hat{p}_3 = 0.421$, $\hat{p}_4 = 0.50$, $\hat{p}_5 = \hat{p}_6 = 0.658$ and the MLE of N is 38 with an estimated standard error of 0.62. This classical model M_t also indicates that nearly all of the mice were caught and a traditional 95% confidence interval indicates that at most one mouse was missed. The fitted values of this model are shown in table 2.3. Because there are six parameters to model six capture probabilities, the fitted number of captures on each occasion matches the data, i.e., $E(n_j) = n_j$ for $j = 1, 2, \ldots, 6$. The other fitted values are based on

$$E(u_j) = N(1 - p_1)(1 - p_2) \cdots (1 - p_{j-1})p_j$$

and

$$E(m_j) = N[1 - (1 - p_1)(1 - p_2) \cdots (1 - p_{j-1})]p_j$$

Under model M_t, the usual chi-squared goodness-of-fit test for multi-nomial counts cannot be computed due to sparseness of the data and relatively many parameters. Although the model is an extension of model M_0, a new chi-square goodness of fit test proposed by Stanley and Burnham (1999, p. 373) shows that this classical model still cannot provide a good fit to the data (chi-squared = 18.67, with 11 degrees of freedom, $p = 0.067$). Therefore, neither of the models M_0 and M_t appears suitable for the data.

Tests of Equal Catchability

With model M_t, a basic assumption is all animals have the same probability of capture at each sampling occasion. Equal catchability may be an unattainable ideal with naturally occurring wild populations. Under some circumstances, however, the effects of violating this assumption can be assessed, and the resulting biases may not be too severe (Carothers 1973a).

If there are only two capture occasions, the equal catchability assumption is not testable because with two occasions there are three parameters and three observable capture frequencies with the multinomial model. Therefore, no degrees of freedom are left for the test of goodness of fit. For more than two occasions, Seber (1982, p.157) described the usual chi-squared goodness-of-fit test for all observed capture frequencies. Large-scale pooling may be necessary to perform this chi-squared test when counts are sparse. A test proposed by Leslie (1958) and further described by Otis et al. (1978, pp. 118–119) and Krebs (1999, section 2.4) has been widely used. However, this test also requires a large number of data. Stanley and Burnham (1999) recently reviewed the available tests of catchability and proposed a more efficient approach.

Causes and Consequences of Assumption Failures

There are two common causes for the failure of the equal catchability assumption:

1. The capture probability of an animal depends on its previous capture history. For example, animals may become either trap happy or trap shy if trapping methods are kept constant. Marked animals that have a higher probability of being captured (e.g., because they are attracted by the bait) on subsequent occasions would be called trap happy. On the other hand, marked animals may become trap shy because of the unpleasant experience of being captured. The recapture

probability on subsequent capture occasions thus becomes lower than that for uncaptured animals.

2. The capture probability is a property of the animal and thus the individual capture probabilities may be heterogeneous across the individuals in a population. Individual capture probabilities may vary with age, sex, body weight, activity, the number of traps near the home range, or other unobservable individual characteristics. For example, studies have shown that for some species the males are consistently more likely to be caught than the females and young animals are more catchable than subadults or adults.

The above two causes are usually mixed and cannot be easily disentangled in a data analysis. Violation of the equal catchability assumption leads to biases for the usual Petersen-Lincoln and Chapman's estimators. For example, if animals exhibit a trap-happy behavioral response or if individual heterogeneity exists and is consistent over sampling occasions, then the animals captured in the first sample are more easily caught in the second sample. Thus, the recapture rate (m_2/n_2) in the second sample tends to be larger than the true proportion of marked animals in the population n_1/N. Then it is expected that $m_2/n_2 > n_1/N$, which yields $N > n_1 n_2/m_2 = \hat{N}_P$. As a result, the Petersen-Lincoln estimator tends to underestimate the true size. Conversely, it tends to overestimate with the trap-shy animals. Similar arguments and conclusions are also valid for more than two capture occasions.

Classical Resolution of Problems Resulting from Assumption Failures

Behavioral response is usually induced by the use of identical trapping methods over the capture occasions. To reduce this effect the use of different trapping methods has been proposed, for example, trapping and then resighting in wildlife studies, or netting and then angling in fishery science. However, this is sometimes infeasible if a number of trapping occasions are used or if available trapping options are limited by costs or logistical constraints.

Some common sources of heterogeneity are observable and can be recorded, e.g., the age, sex, or other characteristics of the individual animals. Such sources can then be eliminated by stratification. That is, analyses are performed within subsets of the data that are homogeneous with regard to the characteristics of animals. However, when there are many stratifying factors, the data in each stratum become sparse. Also, when heterogeneity is due to unobservable or inherent attributes, then residual heterogeneity may still exist even within strata.

2.4 Limitations of Early Models and the Motivation for More General Models

The classical model M_t assumes that all animals have the same probabilities of capture on any particular occasion, though the probability may vary from one occasion to another. The usual estimators under classic models often are either negatively biased (in the trap-happy cases or when heterogeneity exists) or positively biased (in the trap-shy cases). Although these estimators may have good precision (a small variance) and produce narrow confidence intervals, these intervals may be too optimistic and do not reveal their bias. Therefore, models assuming equal catchability may often result in misleading interpretation of the data. In some cases biases resulting from assumption violations may be small and have little impact on study conclusions. It is critical for the investigator, therefore, to be familiar enough with the data and the study situation to interpret data and results of analyses in light of necessary caveats.

Although different sampling methods or stratified analyses have been suggested to minimize trap-response effects or heterogeneity, these suggestions

Figure 2.2. Fin-clipping Dolly Varden (*Salvelinus malma malma*) to test retention of Floy Anchor Tag (center of photo), Buskin River, Kodiak Island, Alaska, 1992. (Photo by Mary Whalen)

have practical limitations. Therefore, more complex models that allow for unequal catchabilities are needed. These are discussed in chapter 4.

2.5 Chapter Summary

- The early history of methods for the estimation of the size of a closed population from capture–recapture data is reviewed, beginning with the work of Graunt about 400 years ago and Laplace about 200 years ago. The form of capture–recapture data is described, with the statistics that are often used to summarize the data for an analysis.
- The common assumptions of capture–recapture models are listed (a constant population size, no loss of marks, correct recording of marks, and animals behave independently).
- The classical Petersen-Lincoln estimator of population size is described and its variance and bias given.
- The Schnabel census with more than two sampling occasions is described with the models M_0 (constant capture probabilities) and M_t (capture probabilities vary with time) used for estimation. Variance equations are provided for both models. Their use is illustrated on a data set involving the captures of deer mice with six sampling occasions.
- Tests are described for the assumption that all animals have the same probability of being captured.
- The causes and consequences of failures of assumptions are discussed, together with ways to avoid the effects of failures. The need for more complicated models as covered in chapter 4 is highlighted.

Three

Classical Open-population Capture–Recapture Models

KENNETH H. POLLOCK AND RUSSELL ALPIZAR-JARA

3.1 Introduction

In the previous chapter closed capture–recapture models were considered for situations where the population size does not change during the study. When open-population models are used, the processes of birth, death, and migration are allowed, and therefore the population size can change during the study. Studies of open populations often cover extended time periods, and the population changes that occur are of great interest to ecologists and managers. The most popular model of this open model class is the Jolly-Seber (JS) model (Jolly 1965; Seber 1965, 1982; Pollock et al. 1990; Schwarz and Seber 1999), which requires that the number of uniquely marked and unmarked animals be recorded on each trapping occasion. Therefore, a complete capture history of each captured animal is available. The model allows estimation of the parameters pertaining to population sizes, survival rates, recruitment numbers, and capture probabilities. However, it is not possible to separate survival from emigration or recruitment from immigration without additional information.

There were early open-population capture–recapture models of importance when they were developed (e.g., Jackson 1939, 1940, 1944, 1948; Fisher and Ford 1947; Leslie and Chitty 1951; Leslie et al. 1953). Entomological research was the motivation for some early model development. Jackson developed early models to study tsetse fly populations in Africa, and R. A. Fisher developed models to help E. B. Ford study a moth population. These models did not handle variability correctly. They were either deterministic or partially deterministic, and underestimated the variances of estimators (Cormack 1968). These models have been superseded by the stochastic Jolly-Seber model and should no longer be used.

Two papers that appeared before the papers of Jolly (1965) and Seber (1965) stand out as crucial to the development of the JS model. In the first, Darroch (1959) presented maximum likelihood estimators for two special cases of the open-population model. These were the additions-only model (only births and immigration allowed) and the deletions-only model (only deaths and emigration allowed). Pollock (2001) in a paper

Figure 3.1. Flipper-tagged sea otter (*Enhydra lutris*), Monterey Bay, California, 1995. (Photo by Steven C. Amstrup)

discussing Seber's contributions to the field notes that it was actually Darroch who introduced Seber to capture–recapture models. That turned out to be a very fortunate personal connection for the development of the field!

The second crucial paper was by Cormack (1964). He considered survival and capture probability estimation for marked birds and derived one component of the likelihood used by Jolly and Seber in their more general model. To recognize Cormack's contribution to development of these models, most scientists now use the term Cormack-Jolly-Seber (CJS) model when referring to the marked animal component of the likelihood function (e.g., Lebreton et al. 1992).

The work of Jolly and Seber has had a tremendous impact in the development of open capture–recapture models. See Seber (1982, p. 196) for an excellent detailed description of the Jolly-Seber model and Pollock et al. (1990) for a comprehensive exposition with several illustrative examples and some model restrictions and extensions. Many new models have followed from this seminal work and these will be presented in later chapters of this book.

In this chapter we discuss the original JS model, its relationship to the CJS model, and some restricted and more generalized versions of the JS model. This is followed by a discussion of age-dependent models, including one proposed by Manly and Parr (1968). General issues of model selection and goodness of fit are considered next, followed by the illustration of the methodology with a comprehensive example.

3.2 The Original Jolly-Seber Model

The following notation consistent with standard usage and table 1.1 of chapter 1 is used in this chapter. Parameters are

p_j = the probability of capture in period j;
ϕ_j = the probability of survival from period j to period $j + 1$;
M_j = the marked population size just before period j;
U_j = the unmarked population size just before period j;
$N_j = M_j + U_j$ = the population size in period j; and
B_j = the birth numbers in the interval from period j to period $j + 1$.

Statistics are

m_j = the number of animals captured at sampling occasion j that are marked;
u_j = the number of animals captured at sampling occasion j that are unmarked;
$n_j = m_j + u_j$ = the total number of animals captured at sampling occasion j;
R_j = the total number of animals captured at sampling occasion j that are released;
r_j = the number of members of the R_j captured again later; and
z_j = the number of members of the marked population not captured at sampling occasion j $(M_j - m_j)$ that are captured again later.

Parameter Estimation

We now present an intuitive derivation of parameter estimates for the standard Jolly-Seber model where capture and survival parameters are allowed to vary in each period. These estimates are based on a maximum likelihood (ML) approach (Seber 1982, p. 198; chapter 1 this volume). While in practice one would obtain the ML estimates using a computer program like JOLLY (Pollock et al. 1990), POPAN (Arnason and Schwarz

1999), or MARK (White and Burnham 1999), we believe these intuitive estimates can help biologists to understand the structure of the (JS) model and also see more clearly why the various assumptions made are so crucial.

All of the necessary estimators rely on the estimation of the marked population size, M_j. Because deletions (deaths and emigration) are possible under open models, M_j must be estimated. This is done by equating the future recapture rates of two distinct groups of animals in the population at time j: (1) those that are already marked but not seen at time j ($M_j - m_j$), and (2) those that are seen and released at time j (R_j). Under the assumption of equal catchability of individuals, the future recapture rates of these two groups should be equivalent. Thus, if z_j and r_j are the members of the $M_j - m_j$ and R_j that are captured again later (at least once), then

$$\frac{z_j}{M_j - m_j} \approx \frac{r_j}{R_j}$$

and

$$\hat{M}_j = m_j + \frac{R_j Z_j}{r_j} \tag{3.1}$$

for $j = 2, \ldots, k - 1$.

The "survival" rate estimator is obtained from the ratio of marked animals present at time $j + 1$ to those present at time j,

$$\hat{\phi}_j = \frac{\hat{M}_{j+1}}{\hat{M}_j + R_j - m_j} \tag{3.2}$$

for $j = 1, \ldots, k - 2$, where the term, $R_j - m_j$, represents the number of newly marked animals released at time j. The survival estimator does not distinguish between losses due to death and permanent emigration without more information. Therefore, this quantity is often called "apparent survival" in the literature.

The capture probability, p_j, is estimated as the ratio of marked animals caught at time j to the number present in the population at time j:

$$\hat{p}_j = \frac{m_j}{\hat{M}_j} \tag{3.3}$$

for $j = 2, 3, \ldots, k - 1$.

The population size for period j can be determined by equating the sample and population ratios of marked to total animals,

$$\frac{m_j}{n_j} \approx \frac{M_j}{N_j}$$

and then, just as with the Petersen-Lincoln estimator,

$$\hat{N}_j = \frac{\hat{M}_j n_j}{m_j} \tag{3.4}$$

or

$$\hat{N}_j = \frac{n_j}{\hat{p}_j}$$

Here, n_j represents the number of animals captured at each sampling occasion, m_j of which are marked.

To estimate birth numbers, the difference in population size at time $j + 1$ and time j is determined, accounting for deaths due to natural mortality $(1 - \phi_j)$ and capture mortality $(n_j - R_j)$:

$$\hat{B}_j = \hat{N}_{j+1} - \hat{\phi}_j [\hat{N}_j - (n_j - R_j)] \tag{3.5}$$

for $j = 2, \ldots, k - 2$. The "birth" number estimator cannot distinguish between individuals entering the population due to recruitment and immigration. This reflects the fact that this estimator for birth is derived purely by subtraction and has nothing to do with the actual processes surrounding birth. Although it has to be considered a poor estimator for recruitment, as opposed to other processes that can bring new animals into a population; it may be possible to separate individuals entering a sample on the basis of their size, sexual maturity, etc. and thereby informally test whether it is indicative of actual births. Negative estimates for this quantity are possible in practice, especially if population sizes and capture probabilities are low (see table 3.3 in the Example in section 3.7 below). Such estimates should be set at zero because there cannot be "negative" births.

The Jolly-Seber model also allows estimation of the probability of an animal being returned to the population in sampling occasion j by R_j/n_j. Its complement estimates the proportion of animals lost on capture.

All of the above estimators are subject to small sample bias, but Seber (1982, p. 204) gives approximately unbiased versions that are as follows:

$$\tilde{N}_j = \frac{\tilde{M}_j(n_j + 1)}{m_j + 1}$$

$$\tilde{\phi}_j = \frac{\tilde{M}_{j+1}}{\hat{M}_j + R_j - m_j}$$

$$\tilde{B}_j = \tilde{N}_{j+1} - \tilde{\phi}_j[\tilde{N}_j - (n_j - R_j)]$$

$$\tilde{p}_j = \frac{m_j}{\tilde{M}_j}$$

and

$$\tilde{M}_j = m_j + \frac{(R_j + 1)z_j}{r_j + 1}$$

The estimated variances and covariances of all these estimators, based on asymptotic theory, are presented by Seber (1982, p. 202) and Pollock et al. (1990). Here we present only the estimated variance for the population size estimate, which is

$$\text{var}(\hat{N}_j \mid N_j) = N_j[N_j - E(n_j)]$$
$$\times \left[\frac{M_j - E(m_j) + R_j}{M_j} \left(\frac{1}{E(r_j)} - \frac{1}{R_j} \right) + \frac{N_j - M_j}{N_j E(m_j)} \right] \quad (3.6)$$

and the total variance for the survival estimate

$$\text{var}(\hat{\phi}_j) = \phi_j^2 \left\{ \frac{[M_{j+1} - E(m_{j+1})][M_{j+1} - E(m_{j+1}) + R_{j+1}]}{M_{j+1}^2} \left[\frac{1}{E(r_{j+1})} - \frac{1}{R_{j+1}} \right] \right.$$
$$\left. + \frac{M_j - E(m_j)}{M_j - E(m_j) + R_j} \left[\frac{1}{E(r_j)} - \frac{1}{R_j} \right] \right\} + \left[\frac{\phi_j(1 - \phi_j)}{M_j - E(m_j) + R_j} \right] \quad (3.7)$$

Because estimates of birth are linked to estimates of survival by simple subtraction (equation 3.5), estimates of survival rates and recruitment numbers can be negatively correlated. Estimates of these parameters should be interpreted with this in mind. Generally, for small samples, estimated variances tend to be smaller than the true variances.

Assumptions

Many assumptions need to be made for the Jolly-Seber model to be valid and for the estimators to be approximately unbiased:

1. every animal alive in the population at a given sample time j has an equal chance (p_j) of being captured in that sample (equal catchability);

2. every marked animal alive in the population at a given sample time j has an equal chance of survival (ϕ_j) until the next sampling occasion (implicitly, this assumption applies to all animals, marked and unmarked, in order to estimate the survival of all animals in the population);

3. marked animals do not lose their marks and marks are not overlooked;

4. sampling periods are short (i.e., effectively instantaneous); and

5. all emigration from the population is permanent.

Properties of Estimators

Pollock et al. (1990) stated that population size estimates are negatively biased (i.e., tend to be too small) by heterogeneity of capture probabilities. This is similar to the situation with closed models that make this assumption (chapter 2 of this volume). Population size may be positively or negatively biased by permanent trap response (chapter 2 of this volume). If animals are trap happy then there is a negative bias, whereas if animals are trap shy then there is a positive bias. Temporary changes in capture rate after marking (i.e., temporary trap response) can be detected and then estimated using the models of Robson (1969), Pollock (1975), and Brownie and Robson (1983).

Survival estimates are relatively robust to heterogeneity of capture probabilities. They also are not affected by trap responses, although measures of precision are affected, depending on the type and magnitude of the violation. With trap-happy animals precision is improved, while for trap-shy animals precision is lessened (Pollock et al. 1990).

Survival probabilities may also be heterogeneous in the population. Pollock et al. (1990) discuss this in some detail. If the same animals tend to have higher or lower survival probabilities consistently from year to year then both survival-rate estimates and population size estimates are positively biased. The situation is still more complex if survival and capture probabilities are both heterogeneous at the same time, a situation that still requires more research. See Pollock and Raveling (1982) and Nichols et al. (1982b) for details in the related setting for band-recovery models.

If marking influences survival, which unfortunately sometimes occurs in practice, then survival rates can be severely underestimated and population sizes will have a positive bias that can also be severe. Some methods of marking fish suffer from this problem (Ricker 1958; Pollock et al. 1990; Hoenig and Pollock, chapter 6 this volume). Temporary reductions in survival rate after marking can be detected and then estimated using the models of Robson (1969), Pollock (1975) and Brownie and Robson (1983).

Any loss or overlooking of tags will result in overestimation of population size in open models, especially if the capture probability is low (McDonald et al. 2003). Survival estimates will be negatively biased by tag loss (Arnason and Mills 1981). Clearly, the choice of tag type (dye, fin clip, ear tag, radio collar, etc.) is important. Double marking schemes can be used to estimate tag loss and adjust for it (Seber 1982, p. 94). Given recent improvements in sighting technology, distinctive natural features of individuals have sometimes been used as natural marks (e.g., see Karanth and Nichols 1998; and Smith et al. 1999).

It is important to emphasize that the positive biases in population sizes, due to tag-loss or tag-induced mortality reported here, are different from the results presented in Pollock et al. (1990, pp. 25–26). They stated that tag loss or tag-induced mortality would not cause a bias in population size estimates. This result was based on large sample arguments, and also assumed that the tag loss rate or tag-induced mortality rate was homogeneous over time. Clearly a realistic model for tag loss would be heterogeneous with some tags lost quickly and some others lost at a slower rate. Under such a model tag loss does cause a serious positive bias unless the capture probability is high as stated previously. A similar argument applies to tag-induced mortality (McDonald et al. 2003).

The assumption of instantaneous sampling periods is important because otherwise animals would have to be assumed to have survived from the midpoint of one sampling interval to the midpoint of the next sampling interval. Animals captured toward the end of the sampling interval would actually have to survive less time to the next sampling interval. This would cause some heterogeneity of survival probabilities but in practice this should not be serious. Sampling periods are never instantaneous, but certainly sampling periods should be kept as short as possible.

Emigration must be permanent for the JS model to be valid. Recall that estimation of the number of marked animals involved using the z_i (animals captured before and after i but not at i), and it is important that these animals be present at time i when they were not captured. Temporary emigration may occur in practice and can be extremely important but this is not discussed until chapter 5.

Pollock et al. (1990, p. 70) present a discussion of precision of the JS estimators. They include some graphs to aid the reader. However, simulation can always be used to study their particular cases in detail. The computer packages MARK and POPAN both include simulation capabilities. Precision is influenced by many factors, such as the number of captured individuals, capture probabilities, and the degree of capture probability heterogeneity. However, in general terms, survival rates and population sizes can be estimated reasonably well, but birth numbers have poor precision unless the capture probabilities are extremely high (Pollock et al. 1990, figure 8.10).

3.3 The Jolly-Seber Likelihood Components

Traditionally the JS model has been based on a maximum likelihood approach to estimation (see chapter 1). Despite the fact that they obtained the same point estimates of the parameters presented earlier, Jolly (1965) and Seber (1965) used different likelihoods. Here the approach of Seber (1982) and Brownie et al. (1986) is used to explain the components of the likelihood. Crosbie and Manly (1985) and Schwarz and Arnason (1996) use a different formulation that handles the birth process in a more elegant manner, as discussed in chapter 5.

The likelihood can be viewed as three conditionally independent components with the overall likelihood the product of these, i.e., $L = L_1 \cdot L_2 \cdot L_2$. The first component, L_1, is a product binomial likelihood that relates the unmarked population and sample sizes at each time to the capture probabilities. The second component, L_2, will not be discussed further here. The only parameters it contains are the probabilities of being lost on capture. The third component, L_3, is the component that contains all the recapture information conditional on the numbers of marked animals released each time. The only parameters contained in this component are the capture and survival probabilities. This is the CJS likelihood that was originally derived by Cormack (1964).

One way to view estimation in the full JS model is that capture and survival probabilities (and marked population sizes) are estimated from L_3. The nuisance probabilities of losses on capture are estimated from L_2, and the unmarked population sizes estimated from L_1. The estimate of the unmarked population takes the form

$$\hat{U}_i = \frac{u_i}{\hat{p}_i}$$

The estimate of population sizes then follows and is the same as that derived earlier in equation 3.4, namely

$$\hat{N}_i = \hat{U}_i + \hat{M}_i = \frac{\hat{M}_i n_i}{m_i}$$

Note that in this formulation the population sizes and birth numbers do not appear in the likelihood directly and are derived parameters.

The Cormack-Jolly-Seber Model

In some cases all components of the likelihood are combined in the general JS model as we just described. However, in other cases the CJS model may be used alone to estimate survival and capture probabilities (Cormack 1964). This CJS model requires information on only the recaptures of the marked animals, and that the marked animals be representative of the population. In some cases the marking process may use a totally different method of capture than the recapture process, as with mark–resight studies. See, for example, Pollock et al. (1990, p. 51) for a description of a mark–resight study on Canada geese. The dependence, in the CJS model, on only recaptures of marked animals means that population size cannot be estimated directly.

The original CJS model was for one group of animals and assumed no age-dependence of survival and capture probabilities. Lebreton et al. (1992) presented a unified approach to extensions of these models that allows modeling of survival and capture (sighting) probabilities as function of time, age, and categorical variables characterizing the individuals. This is discussed in chapter 5.

3.4 Restrictions and Generalizations of the Jolly-Seber Model

Various restricted versions of the general JS model have the substantial advantage of reducing the numbers of parameters to be estimated. The deaths-only model may sometimes be useful if there is no immigration and recruitment is not occurring. Similarly the births-only model may sometimes be useful if there is no mortality or emigration (Darroch 1959). Both restricted versions of the general model are implemented in the computer program JOLLY. Versions of the general model with the restrictions of constant survival and/or capture probabilities were developed by Jolly (1982) and Brownie et al. (1986), and they have also been implemented in JOLLY. They will be illustrated by an example later.

Generalizations of the JS model that allow for temporary effects on survival and capture rates have been considered. Temporary reductions in survival rate after marking and temporary changes in capture probability can be detected and then estimated using the models of Robson (1969), Pollock (1975), and Brownie and Robson (1983). Some of these models have also been implemented in JOLLY and they can easily be implemented in MARK.

3.5 Age-dependent Models

Generalizations of the JS model such as those allowing age dependence are also possible. Early work was by Manly and Parr (1968), whose model also illustrates some features of the JS model. As we discussed earlier, the estimator of population size in the JS model can be written

$$\hat{N}_i = \frac{n_i}{\hat{p}_i}$$

with the estimator of capture probability given by

$$\hat{p}_i = \frac{m_i}{\hat{M}_i}$$

Manly and Parr (1968) realized that one could estimate p_i in a manner that was robust to heterogeneity of survival rates due to age and came up with a different estimator. They defined C_i as the class of marked animals known to be alive at time i because they were captured before and after i, and they defined c_i as the members of that class C_i that are captured at i. Therefore, an estimate of p_i is

$$\tilde{p}_i = \frac{c_i}{C_i}$$

and hence

$$\tilde{N}_i = \frac{n_i}{\tilde{p}_i} = \frac{n_i C_i}{c_i}$$

A Chapman modification to reduce bias is then

$$\tilde{N}_i = \frac{(n_i + 1)(C_i + 1)}{c_i + 1} - 1$$

Also survival rate and birth numbers can be estimated as described in Seber (1982, p. 236). We believe all these estimators will have larger standard errors than the Jolly-Seber estimators because they use a smaller class of marked animals (C_i) as the basis of their estimation rather than the total marked population (M_i). However, this model has not been much studied, and Seber (1982, p. 234) presents only an approximate variance for the population size estimators. In section 3.7 we present a small example and contrast the Manly-Parr estimates with the JS estimates.

There has been much work on age-dependent models. A focus of particular interest has been the situation when more than one identifiable age class is marked (Pollock 1981a; Pollock and Mann 1983; Stokes 1984; Loery et al. 1987; Pollock et al. 1990). This research led to the computer program JOLLYAGE. Recently MARK (White and Burnham 1999) has made many of these analyses very easy to perform, as considered further in chapter 5.

3.6 Goodness-of-Fit and Model Selection Issues

There are two aspects to assessing whether the best model for an open capture–recapture model has been chosen. There is the overall goodness of fit of a model, and the decision about which of a series of related models is the best. Some of this material has been discussed in chapter 1, and only a very brief summary is presented here.

Assumption violations may mean that a model does not fit the capture–recapture data available. Seber (1982, p. 223) suggested the standard chi-squared test based on comparing observed and expected values for assessing the general JS model. Pollock et al. (1985) derived two related tests based on minimal sufficient statistics. Each chi-squared test had two contingency tables based on conditional arguments and sufficient statistics for the general JS model. Brownie et al. (1986) extended these goodness fit tests to the restricted models B (constant survival) and D (constant survival and constant capture). All of these tests are implemented in JOLLY and are used in an example that follows.

Traditionally likelihood ratio-type tests have been used to choose which model to fit from a nested set (chapter 1). Brownie et al. (1986) used a related approach based on conditional sufficient statistics to compare models, and these have been implemented in JOLLY. There is an example in section 3.7. The modern approach to testing between models is to use the AIC procedure described briefly in chapter 1. It allows the comparison of nonnested models, and puts a penalty on fitting a model with too many parameters. The AIC method is used with the CJS example using MARK presented below.

TABLE 3.1
Alligator between-year capture-history data collected by Fuller (1981) for a
population at Lake Ellis Simon, North Carolina, from 1976 to 1979

1976	1977	1978	1979
$n_1 = 20$	$n_2 = 51$	$n_3 = 41$	$n_4 = 25$
$R_1 = 17$	$R_2 = 49$	$R_3 = 37$	$R_4 = 25$
	$X_{11} = 9$	$X_{111} = 4$	$X_{1111} = 1$
	$X_{01} = 42$	$X_{101} = 1$	$X_{1011} = 1$
		$X_{011} = 18$	$X_{0111} = 6$
		$X_{001} = 18$	$X_{0011} = 5$
			$X_{1001} = 1$
			$X_{1101} = 2$
			$X_{0101} = 4$
			$X_{0001} = 5$

Note. n_i = the number of alligators captured in year i (e.g., $n_1 = 20$ is the number of alligators captured in year 1976); R_i = the number of alligators release in year i (e.g., $R_1 = 17$ is the
number of alligators released in year 1976); X_{ijkl} = the number of alligators with capture history $ijkl$, where $ijkl$ denotes years 1976, 1977, 1978, 1979 and where 1 denotes capture and
0 indicates that the animal was not captured (e.g., $X_{111} = 4$ is the number of alligators first
captured in year 1976 that are recaptured in year 1977 and then again in year 1978).

3.7 Examples

A Jolly-Seber Example

To illustrate the calculation of statistics and parameter estimation we use
data from a capture–recapture study on a population of the American alligator (*Alligator mississippiensis*) at Lake Ellis Simon, North Carolina,
between 1976 and 1979 (Fuller 1981). This data set has been previously
analyzed to illustrate the advantages of Pollock's robust design in capture–
recapture experiments, and is directly taken from Pollock (1982, p. 756)
and Pollock et al. (1990, p. 60). This data set is small enough so that it is
easy to handle the computations with a calculator. The estimation of parameters with larger data sets is practical only with the use of computer
programs that are readily available, such as JOLLY, POPAN, or MARK.
The capture-history data are presented in table 3.1 and the capture-
history matrix in table 3.2. The data are analyzed first by doing the direct calculations for the full model, and then with the program JOLLY
using the data as presented in table 3.2. There was an error in the data as
originally published, and $X_{1001} = 1$ and not 0.

By definition m_i = the number of marked animals captured in the ith
sample ($i = 2, \ldots, $ k). Also, from table 3.1,

TABLE 3.2
Capture-history matrix for the Jolly-Seber model (JOLLY) and the Cormack-
Jolly-Seber model (MARK)

1976	1977	1978	1979	Number of alligators
1	1	1	1	1
1	0	1	1	1
0	1	1	1	6[a]
0	0	1	1	5
1	0	0	1	1
1	1	0	1	2
0	1	0	1	4
0	0	0	1	5
1	1	1	0	3
0	1	1	0	12
0	0	1	0	9
0	0	2[b]	0	4
1	1	0	0	3
0	1	0	0	18
0	2[b]	0	0	2
1	0	0	0	6
2[b]	0	0	0	3

[a] Reflects 6 identical lines in the real capture history.
[b] These animals were lost on capture and therefore not released. There were no alligators with capture history (1010).

$m_1 = 0$ by definition;
$m_2 = X_{11} = 9$;
$m_3 = X_{111} + X_{101} + X_{011} = 4 + 1 + 18 = 23$;
$m_4 = X_{1111} + X_{1011} + X_{0111} + X_{0011} + X_{1001} + X_{1101} + X_{0101}$
$= 1 + 1 + 6 + 5 + 1 + 2 + 4 = 20$;
r_i = the number of the R_i animals released at i that are captured again $(i = 1, \ldots, k-1)$;
$r_1 = X_{1111} + X_{1011} + X_{1001} + X_{1101} + X_{1110} + X_{1010} + X_{1100}$
$= 1 + 1 + 1 + 2 + 3 + 0 + 3 = 11$;
$r_2 = X_{1111} + X_{0111} + X_{1101} + X_{1110} + X_{0110} + X_{1101}$
$= 1 + 6 + 2 + 4 + 3 + 12 = 28$;
$r_3 = X_{1111} + X_{1011} + X_{0111} + X_{0011} = 1 + 1 + 6 + 5 = 13$;
$r_4 = 0$ by definition;
z_i = the number of animals captured before i, not captured at i that are capture again later $(i = 2, \ldots, k-1)$;
$z_1 = z_4 = 0$ by definition;
$z_2 = X_{1011} + X_{1001} + X_{1010} = 1 + 1 + 0 = 2$; and
$z_3 = X_{1001} + X_{1101} + X_{0101} = 1 + 2 + 4 = 7$.

Based on these summary statistics the bias adjusted parameter estimates can be obtained. Bias-adjusted estimates can also be computed using program JOLLY with model A, and the standard errors reported here are from the output from this program.

First note that we can estimate the marked population sizes for years 2 and 3, so that

$$\tilde{M}_2 = m_2 + \frac{(R_2 + 1)z_2}{r_2 + 1} = 9 + \frac{(49 + 1)2}{28 + 1} = 12.45 \quad (SE = 1.66)$$

and

$$\tilde{M}_3 = m_3 + \frac{(R_3 + 1)z_3}{r_3 + 1} = 23 + \frac{(37 + 1)7}{13 + 1} = 42 \quad (SE = 7.29)$$

Also, survival rates for year 1–2 and year 2–3 are

$$\tilde{\phi}_1 = \frac{\tilde{M}_2}{R_1} = \frac{12.45}{17} = 0.73 \quad (SE = 0.15)$$

and

$$\tilde{\phi}_2 = \frac{\tilde{M}_3}{\tilde{M}_2 - m_2 + R_2} = \frac{42}{12.45 - 9 + 49} = 0.80 \quad (SE = 0.15)$$

Notice that the estimates of survival are quite similar. The relative precision of an estimate (the coefficient of variation, CV) is the standard error divided by the estimate itself. The CVs in the present case 0.15/0.73 and 0.15/0.80, respectively, or 20.5 and 18.7%, respectively. This level of precision is quite reasonable in wildlife science.

It is possible to estimate the population sizes in years 2 and 3, so that

$$\tilde{N}_2 = \frac{(n_2 + 1)\tilde{M}_2}{m_2 + 1} = \frac{(51 + 1)(12.45)}{9 + 1} = 64.7 \approx 65 \quad (SE = 12.3)$$

and

$$\tilde{N}_3 = \frac{(n_3 + 1)\tilde{M}_3}{m_3 + 1} = \frac{(41 + 1)(42)}{23 + 1} = 73.5 \approx 74 \quad (SE = 14.9)$$

The relative precisions for the population size estimates are similar to those for the survival estimates, at approximately 20%.

The capture probabilities for years 2 and 3 are

$$\tilde{p}_2 = \frac{m_2}{\tilde{M}_2} = \frac{9}{12.45} = 0.72 \quad (SE = 0.16)$$

and

$$\tilde{p}_3 = \frac{m_3}{\tilde{M}_3} = \frac{23}{42} = 0.55 \quad (SE = 0.12)$$

For this small example with four sampling periods it is possible to estimate only one set of birth numbers, for the period between years 2 and 3:

$$\tilde{B}_2 = \tilde{N}_3 - \tilde{\phi}_2(\tilde{N}_2 - n_2 + R_2) = 73.5 - 0.8(64.7 - 51 + 49)$$
$$= 73.5 - 50.2 = 23.3 \quad (SE = 13.1)$$

This estimate has poor relative precision (13.1/23.3, or 56.2%). Poor precision for estimates of birth numbers is common in practice (Pollock et al. 1990, p. 71).

Restricted Parameter Model Output from Program JOLLY

Jolly (1982) showed that when restricted models (with the assumptions of constant survival and/or capture probabilty over time) are reasonable, then there can be a large gain in the precision of estimators compared to the standard JS estimators because these parsimonious models have many fewer parameters. Of course, if the assumptions of constancy of survival or capture probability are not valid then bias will be introduced. Program JOLLY offers features such as maximum likelihood parameter estimates, goodness-of-fit tests, and methods to aid in model choice based on likelihood ratio-type tests, but the newer AIC method of model choice is not available. The calculations above suggest that the survival rates for year 1–2 and year 2–3 are very similar. Hence, it may be useful to test the assumption of equal survival rates to reduce the number of model parameters and gain some precision in the estimates. Consider two restricted versions of the original JS model that are implemented in the program JOLLY. These are model B, which assumes that survival probabilities are constant over the whole experiment (i.e., $\phi_1 = \phi_2 = \phi_3$), but capture probabilities may vary from year to year, and model D, which assumes that

TABLE 3.3

Estimates and approximate standard errors (in parentheses) under the Jolly-Seber model (A), the constant survival model (B), and the constant survival and capture model (D) for an alligator population at Lake Ellis Simon, North Carolina, from 1976 to 1979

Model/year	$\hat{\phi}_i$	$\hat{p}_i$	$\hat{M}_i$	$\hat{N}_i$	$\hat{B}_i$
		Original Jolly-Seber model (model A)			
1976	0.73 (0.15)	—	—	—	—
1977	0.80 (0.15)	0.72 (0.16)	12.45 (1.7)	64.7 (12.3)	23.3 (13.1)
1978	—	0.55 (0.12)	42.0 (7.3)	73.5 (14.9)	—
Mean	0.77 (0.10)	0.64 (0.20)			
		Constant survival model (model B)			
1977		0.70 (0.15)	12.7 (1.5)	73.0 (14.0)	18.2 (12.7)
1978	0.78[a] (0.09)	0.56 (0.11)	41.9 (5.7)	74.2 (10.9)	0[b] (7.2)
1979		0.47 (0.13)	43.0 (9.7)	53.7 (12.2)	—
		Constant survival and capture model (model D)			
1977			13.5 (1.9)	85.6 (13.5)	9.1 (8.9)
1978	0.73[a] (0.07)	0.58[a] (0.09)	39.5 (4.7)	70.4 (9.7)	0[b] (5.3)
1979			34.3 (6.9)	42.9 (8.2)	—

[a] These estimates correspond to all years in the study given the constancy of parameters over time assumptions.

[b] These estimates were negative even though this is biologically impossible. They were set to 0.

both the survival and capture probabilities are constant over the whole experiment (i.e., $\phi_1 = \phi_2 = \phi_3$ and $p_2 = p_3 = p_4$). Table 3.3 shows the parameter estimates for the three models A, B, and D.

Table 3.4 presents a summary of omnibus goodness-of-fit tests and tests comparing models. This can be used to aid in model selection. Small significant p values indicate a lack of fit, so that the three models all fit the data reasonably well. However, a nonsignificant goodness-of-fit test does not necessarily guarantee that all the model assumptions are met, because these tests may have low power. The other likelihood ratio-type tests are used for comparing two models, with the simpler model being the null and the more complex model the alternative. Small p values suggest that the null model should be rejected. Here none of the likelihood ratio-type tests are significant at any reasonable level. This suggests, following the principle of parsimony (choose an acceptable model with the smallest number of parameters), that the model D estimates are the best as a summary of the population.

Model D seems rather restrictive biologically and the time-dependent estimates of capture in model A do appear to vary quite a lot. This is a small example and another more complex model may have been chosen

TABLE 3.4

Tests for the Jolly-Seber model (A), the constant survival model (B), and the constant survival and capture model (D) for an alligator population at Lake Ellis Simon, North Carolina, from 1976 to 1979

Test	χ^2	df	P
Goodness of fit to model A	0.41	2	0.814
Goodness of fit to model B	0.55	3	0.908
Goodness of fit to model D	2.06	5	0.841
Model D vs. model A	1.64	3	0.649
Model B vs. model A	0.14	1	0.712
Model D vs. model B	1.49	2	0.474

if there had been more animals marked and recaptured. Note that model B and model D allow the estimation of additional parameters when compared to the full JS model. Negative birth numbers estimates are possible in these models and they occur for both model B and model D. These have been set to zero, as clearly such estimates are biologically impossible.

In 1978 it is estimated that there were about 70 animals (SE = 9.7) from model D. In that year it was possible to carry out a closed-population analysis and it was estimated that there were about 140 animals (SE = 28.5) using model M_h, the heterogeneity model (Pollock 1982). There is a large discrepancy in the estimates; and even though the standard errors are large there appears to be something else important going on. It is suspected that there was substantial heterogenity of capture probablities due to the large size range of the alligators considered, and that this caused a negative bias in the population size estimates. The so-called robust design allows for this problem and will be discussed in chapter 5. Notice that the goodness-of-fit tests did not pick up this important problem, probably because this is a small population with only four years of data, and therefore the tests have low power (Pollock et al. 1990).

Cormack-Jolly-Seber Component Example in MARK

The CJS model allows the estimation of capture and survival rate estimates, based on the recapture component of the likelihood, using the capture history data in table 3.2 and the recaptures-only option in MARK. The data format had to be modified slightly from that used for JOLLY, but the data were the same. The first step was to use the model-selection procedure to decide on a parsimonious model using the AIC criteria, as discussed in chapter 1. Four models were considered, where survival was either varying over time ($\phi(t)$) or constant over time ($\phi(\cdot)$) and capture

TABLE 3.5
Program MARK AIC model comparisons for an alligator population at Lake El-lis Simon, North Carolina, from 1976 to 1979

Model	AICc	Delta AICc	Number of parameters
$\phi(\cdot)\,p(\cdot)$	185.83	0.00	2
$\phi(\cdot)\,p(t)$	188.54	2.71	4
$\phi(t)\,p(\cdot)$	189.21	3.37	4
$\phi(t)\,p(t)$	190.62	4.78	5

Note. Four models are fitted using varying assumptions about constancy of survival and capture probabilities over time.

probability was also either varying over time $(p(t))$ or constant over time $(p(\cdot))$. The simplest model has two parameters $(\phi,\,p)$ and the full model has five parameters $(\phi_1,\,\phi_2,\,p_2,\,p_3,\,\phi_3 p_4)$. It is not possible to separately estimate the last survival and capture probability parameters. The other two models have 4 parameters (3 capture probabilities and 1 survival probability, or 3 survival probabilities and 1 capture probability). In this case the winner in terms of minimum AIC is the simplest model with both capture and survival constant over time (table 3.5). This is the same result as was obtained with the program JOLLY but that older program does not use the AIC criteria.

The estimates of constant survival and capture probabilty were $\hat{\phi} = 0.73$ (SE = 0.08), and $\hat{p} = 0.58$ (SE = 0.10), respectively. These estimates are the same as those reported in table 3.3 for model D using program JOLLY, but there are small differences in the standard error estimates due to the different algorithms used for the calculations.

Manly-Parr Example

Table 3.6 shows Manly-Parr estimates for the alligator data, based on the equations in section 3.5. The calculations at the base of the Manly-Parr method for time 2 are the total number of marked animals known to be alive at time 2 (C_2), and the number of those captured (c_2). These are

$$C_2 = X_{111} + X_{101} + X_{1101} + X_{1001} = 4 + 1 + 2 + 1 = 8$$

and

$$c_2 = X_{111} + X_{1101} = 4 + 2 = 6$$

Similar calculations for time 3 give $C_3 = 15$ and $c_3 = 8$.

TABLE 3.6
Comparison of Manly-Parr and Jolly-Seber parameter estimates for the alligator data at Lake Ellis Simon, North Carolina, from 1976 to 1979

Parameter	Manly-Parr estimate	Jolly-Seber estimate
N_2	66 (13.2)	65 (12.3)
N_3	74 (16.9)	74 (14.9)
p_2	0.75	0.72
p_3	0.53	0.55
ϕ_1	0.71	0.73
ϕ_2	0.85	0.80
B_2	19.6	23.3

The estimates of the Manly-Parr method for population size, capture probability, survival rate, and birth number can be obtained using the equations given in the previous section. These estimates allow for age-dependent survival rates. Table 3.6 also compares these Manly-Parr estimates to the JS estimates calculated previously, and in this case all of the estimates are very similar. The standard errors for the population size estimates are slightly larger for the Manly-Parr method, as expected because it uses the data in a less efficient way.

3.8 Conclusions

The Jolly-Seber capture–recapture model is important because the estimation of survival, birth numbers, and population sizes is done in one analysis, based on a well-defined stochastic model. However, one major problem with these models is that unlike the closed capture–recapture models they do not allow for unequal catchability of individual animals due to heterogeneity or trap response. In chapter 5, a solution to this problem is given, together with the development of many other new methods based on the original JS model.

3.9 Chapter Summary

- The early history of models for open-population capture–recapture data is reviewed, leading up to the stochastic models of Darroch, Cormack, Jolly, and Seber.
- The Jolly-Seber model is described, with intuitive explanations for the estimators of survival probabilities, capture probabilities, population sizes, and birth numbers. Equations for the variances of

Figure 3.2. Jerry Hupp prepares to release a Canada goose (*Branta canadensis*) after surgical implantation of a VHF radio-transmitter, Anchorage, Alaska, 1999. Also note the double leg bands. (Photo by John Pearce)

the estimators of population size and survival probabilities are provided. The assumptions needed for the use of the model are discussed, together with the general properties of the estimators.

• The components of the likelihood function for the Jolly-Seber model are defined. Components of the Cormack-Jolly-Seber model, which is conditional on the first releases of marked animals, are also defined.

- Restrictions and generalizations of the Jolly-Seber model are discussed.
- The models of Manly and Parr and others allowing for parameters to vary with the age of animals are discussed.
- Tests for goodness of fit and model selection are briefly reviewed.
- The Jolly-Seber method is illustrated using data from a study of the American alligator. The same data are then analyzed using models that assume that the probabilities of capture and/or the probabilities of survival are constant.
- The use of the Cormack-Jolly-Seber model for the same data is illustrated using the program MARK. The Manly and Parr estimates are also shown.

Four

Modern Closed-population Capture–Recapture Models

ANNE CHAO AND RICHARD M. HUGGINS

4.1 Introduction

The models discussed in chapter 2 for estimating the size of a closed population assume that on each capture occasion all animals have the same capture probability. If this assumption is violated, the estimators of chapter 2 may be biased and the estimated standard errors may be too small, resulting in artificially narrow confidence intervals and possibly misleading interpretations of the data. The bias could be severe when capture probabilities vary greatly among animals. To take account of factors that may potentially affect the capture probabilities, more realistic models are needed. In the present chapter models are classified as being one of two types: discrete-time and continuous-time models, discussed respectively in sections 4.2 and 4.3. The incorporation of covariates in analyses is also considered. Computing considerations are discussed in section 4.4.

A number of approaches have been proposed to relax the assumption of equal catchability. The theses of Burnham (1972) and Pollock (1974) were two pioneering works, while Otis et al. (1978) provided a thorough description of multiple-occasion models, upon which modern developments have been based. Otis et al. gave a hierarchical development from equal catchability models to models that incorporate time variation, behavioral response, and heterogeneity of the capture probabilities. Section 4.2 summarizes the strengths, weaknesses, and applications of those models. Since 1980, a wide range of statistical methodologies with and without covariates has been developed to estimate population size, based on the framework of Otis et al. (1978). See Schwarz and Seber (1999), Buckland et al. (2000), Pollock (2000), Seber (2001), and Chao (2001) for recent reviews and general theory.

4.2 Discrete-time Models with Unequal Catchabilities

In a typical discrete-time model, trappings are conducted on a certain number of distinct occasions. Usually each trapping day or trapping night is

Figure 4.1. Mist-netting Bristle-thighed Curlews (*Numenius tahitiensis*) on the Seward Peninsula, spring 1993. (Photo by Robert Gill)

treated as an occasion, and a group of animals are caught on each occasion. However, the exact capture time of each animal in a group is unknown, so that information about the order of captures is not available, and recaptures within the capture occasion are not allowed. The latter can occur if animals are released as soon as possible after their capture, but these recaptures are usually ignored.

The data shown in table 2.1 provide an example for a discrete-time experiment conducted over six occasions. During the first night 15 mice are caught, but the specific capture times for these 15 mice are not known. The maximum capture frequency for any animal was six in that experiment.

Models and Estimators in Otis et al. (1978)

Otis et al. (1978) considered the following three sources of variations in capture probabilities:

- Time effects, where capture probabilities may vary due to environmental variables (e.g., temperature, time of day, rainfall, or humidity) or sampling effort. However, covariates to explain the variation are

not available. Models that include only these time effects are discussed in chapter 2.
- Behavioral responses to capture (e.g., trap shyness).
- Individual heterogeneity due to observable factors (e.g., sex, age, or body weight) or unobservable inherent characteristics.

Based on these three sources of variation, Otis et al. (1978) and White et al. (1982) considered all the possible combinations of sources, and formulated eight models, including the starting null model M_0. They also developed a computer program called CAPTURE (Rexstad and Burnham 1991) to calculate estimators for each model. The advantages and weaknesses of each model and the associated estimators are discussed below, with the data of table 2.1 used to illustrate the application of each model. To describe the models let P_{ij} denote the probability the ith individual is captured on the jth occasion. The subscripts t, b, h, on M denote time variation, behavioral response, and heterogeneity, respectively. The subscript 0 denotes the null model.

MODEL M_0: $P_{ij} = p$

This simplest model is used as a starting model but has limited use in practice. It is the first classical model discussed in chapter 2. As previously stated, there are only two parameters, N and p, in the model, and the likelihood is given by equation 2.4. For the data in table 2.1, minus twice the maximized log-likelihood $-2\ln(L_0) = 109.5$ and the Akaike's information criterion (AIC) is 113.5, where the subscript on L_0 refers to the underlying model. Note that throughout this chapter the common term $\prod a_h!$ in the denominator of likelihood functions is excluded in the computation of likelihoods and AIC values.

MODEL M_t: $P_{ij} = p_j$

This model with time-varying capture probability is the second classical model of chapter 2. There are $k + 1$ parameters and the likelihood is given in equation 2.5. When this model is fitted to the mouse data in table 2.1 the maximized log-likelihood function gives $-2\ln(L_t) = 99.7$ and AIC = 113.7. As discussed later in section 2.3, neither model M_0 nor model M_t appears to fit the data well.

MODEL M_b: $P_{ij} = p$ UNTIL THE FIRST CAPTURE AND
$P_{ij} = c$ FOR ANY RECAPTURES

Empirical studies (e.g., Bailey 1969) have provided evidence that mice, voles, and other small mammals, as well as other species, exhibit trap response to capture, especially the first capture. This model with behavioral response requires that all animals have the same behavioral response to initial capture. On any fixed occasion, the capture probability for animals not previously caught is p, whereas the probability is c ($c \neq p$) for any previously captured animal. If $p > c$, then this gives the trap-shy case; while if $p < c$, then it gives the trap-happy case. Because individual heterogeneity is not considered, all animals are assumed to behave in the same fashion with regard to their trap response.

With model M_b the probability for any animal having the capture history 111111 shown in table 2.1 is pc^5, the probability for any animal having capture history 100111 is $p(1 - c)^2 c^3, \ldots$, and the probability an animal is not captured is $(1 - p)^6$. When all possible capture frequencies and the corresponding counts are considered, the likelihood for M_b based on a multinomial distribution can be obtained and is given by Otis et al. (1978, p. 107).

There are three parameters (N, p, c) for the model, and three statistics are sufficient for calculating maximum likelihood estimates (MLEs). These statistics are $(M_{k+1}, m_., M_.)$, where $m_. = m_1 + m_2 + \cdots + m_k$ and $M_. = M_1 + M_2 + \cdots + M_k$. The MLEs were derived by Moran (1951) and Zippin (1956, 1958). It is found that recapture information $(m_.)$ is used only for the estimation of c, while only the first capture information is used in estimating the population size. The first-caught individuals on each occasion can be conceptually regarded as removed by being marked. Therefore, from the standpoint of estimating the population size, removal by marking is equivalent to permanent removal. Thus, model M_b is statistically equivalent to a removal model.

Model M_b is a simple model that uses only two parameters, p and c, to model behavioral response. However, it has some weaknesses: (1) it must be assumed that all animals exhibit the same behavioral response to capture; (2) as stated above, only the first captures are used for the estimation of population size; and (3) the number of first captures must satisfy some conditions to assure the existence of the MLE (e.g., $u_1 > u_2$ for two occasions, and $u_1 > u_3$ for three occasions (Seber 1982, p. 312)). Generally, the numbers of first captures must decrease with time for MLEs to exist.

For the data in table 2.1, the MLE from CAPTURE turns out to be $\hat{N} = 41$ with a standard error of 3.1, while $\hat{p} = 0.34$, $\hat{c} = 0.61$, suggesting that the animals became trap happy on their first capture. Otis et al. (1978, p. 32) found that model M_b fits the data well based on an overall goodness-of-fit test.

Under model M_b, $-2\ln(L_b) = 98.0$ and AIC $= 104.0$. Comparing this with model M_0, model M_b provides a significantly better fit as twice the difference of the two log-likelihood functions is $2\ln(L_b) - 2\ln(L_0) = 11.5$, which is very significantly large for a chi-squared distribution with 1 degree of freedom ($p < 0.001$). A similar comparison between models M_0 and M_t shows that these two models are not significantly different, which is consistent with the finding from a model selection procedure of Otis et al. (1978, p. 32).

MODEL M_{tb}: $P_{ij} = p_j$ UNTIL THE FIRST CAPTURE AND $P_{ij} = c_j$ FOR ANY RECAPTURES

This is a natural extension of model M_t to incorporate behavioral response. There are a total of $2k$ parameters (N, p_1, p_2, . . . , p_k, c_2, c_3, . . . , c_k), but as shown by Otis et al. (1978, p. 38) the statistics needed to calculate the likelihood function are the $2k - 1$ values (u_1, u_2, . . . , u_k, m_2, m_3, . . . , m_k). Consequently, the model is nonidentifiable (with more parameters than data values) unless some restrictions or assumptions are made. Despite the nonidentifiability, model M_{tb} has been selected as the most likely model for estimating some quail and mice populations (Pollock et al. 1990, pp. 15–17; Otis et al. 1978, p. 94). This has motivated further research on setting reasonable restrictions to remove the nonidentification, as considered later in this chapter.

MODEL M_h: $P_{ij} = p_i$

Model M_h assumes that the ith animal has its own unique capture probability, p_i, which is constant over all trapping occasions. As heterogeneity is expected in almost all natural populations, this model is useful for many species such as rabbits (Edwards and Eberhardt 1967), chipmunks (Mares et al. 1981), skunks (Greenwood et al. 1985), and grizzly bears (Boyce et al. 2001; Keating et al. 2002). Many previous studies have confirmed that the usual estimators based on the classical equal catchability assumption are negatively biased by heterogeneity of capture probabilities.

The main difficulty with this model lies with the large numbers of nuisance parameters, i.e., (p_1, p_2, . . . , p_N). Without any assumptions or restrictions on these the likelihood is difficult to handle using classical inference procedures. One plausible approach is to assume a parametric distribution for p_i such as the beta distribution, so that all the p_i values are replaced by the two parameters of the beta distribution. However, Burnham (Otis et al. 1978, p. 33) found that the MLE for this parametric setup were unsatisfactory. This motivated him to develop a nonparametric approach. Under the assumption that (p_1, p_2, . . . , p_N) are a random sample from any unknown distribution, Burnham and Overton (1978) showed that the capture frequencies ($f_1, f_2, . . . , f_k$)

are all that is needed for estimating the population size, where f_j denotes the number of animals captured exactly j times on occasions 1, 2, . . . , k.

Burnham and Overton (1978) proposed the use of jackknife estimators up to the fifth order for estimating population size. Jackknife estimators were developed as a general technique to reduce the bias of a biased estimator. Here the biased estimator is the number of animals observed. The basic idea with the jth-order jackknife method is to consider subsets of the data by successively deleting j occasions from the original data. An appealing property of the jackknife approach is that the final estimates have closed forms. For example, the first-order and the second-order jackknife estimators turn out to be

$$\hat{N}_{j1} = M_{k+1} + \left[\frac{k-1}{k}\right] f_1 \tag{4.1a}$$

$$\hat{N}_{j2} = M_{k+1} + \left[\frac{2k-3}{k}\right] f_1 - \left[\frac{(k-2)^2}{k(k-1)}\right] f_2 \tag{4.1b}$$

Higher orders of the jackknife estimators are given in Otis et al. (1978, p. 109). Burnham and Overton (1978) also proposed a sequential testing procedure to select the best order. They recommended an interpolated jackknife estimator. All jackknife estimators are linear functions of capture frequencies and approximate variances can be derived.

Although Cormack (1989, p. 404) pointed out that there is no theoretical advantage to using this approach, the jackknife has performed well in simulations. Based on various simulation results, Otis et al. (1978, p. 34) commented that the jackknife estimator has a tolerable bias if the number of trapping occasions is sufficiently large (say greater than 5) and if a negligible number of animals are untrappable. As $f_1 \le M_{k+1}$, it is seen from (4.1a) that the first-order jackknife estimate is always less than $2M_{k+1}$. Therefore, it usually underestimates the true population size when the mean capture probabilities are small, as shown in Chao (1988). On the other hand, if nearly all animals are captured, the jackknife methods tend to overestimate the population size. The jackknife methodology does not provide a measure to quantify the degree of heterogeneity among the animals.

Applying the jackknife technique to the data in table 2.1 by substituting $M_{k+1} = 38$, $k = 6$, $f_1 = 9$, and $f_2 = 6$ into (4.1) gives $\hat{N}_{j1} = 45.5$ (standard error 3.7) and $\hat{N}_{j2} = 48.3$ (standard error 5.7). A sequential test applied to M_{k+1} and successive orders of jackknife estimators results in an interpolated jackknife of 44.0 (standard error 4.0).

Since the proposal of the jackknife estimator, a number of researchers have presented various alternative approaches and compared their relative merits. The advances are quite substantial and some of them will be introduced later.

MODEL M_{bh}: $P_{ij} = p_i$ UNTIL FIRST CAPTURE AND $P_{ij} = c_i$ FOR ANY RECAPTURE

If model M_h is generalized to allow the individual capture probability to depend on previous capture history, it gives model M_{bh}. This model is equivalent to a heterogeneous removal model in which each animal is removed with a different probability. Thus, it also finds wide applications in removal and catch-effort studies, especially in fishery science.

Model M_{bh} has some of the disadvantages of models M_b and M_h. Like model M_b, only the first captures carry information about the population size. Like model M_h, the individual heterogeneity effects cause theoretical difficulties. For this model, Otis et al. (1978) proposed a generalized removal method. This is based on the idea that animals with larger removal probabilities tend to be removed in earlier occasions than animals with small removal probabilities. Conditional on those animals not previously removed, the average removal probabilities of the k occasions are expected to be decreasing. For a homogeneous population, these conditional removal probabilities are constant.

The generalized removal method first considers fitting a simple model with all average removal probabilities equal. If the simple model fits the data well, then the model is reduced to model M_b. In this case, the inferences are identical to those of model M_b. If this model does not fit, then consider a model where the removal probability on the first occasion is larger than the others. In other words, it is basically assumed that after u_1 animals have been removed, the remaining $N - u_1$ animals have identical probabilities of being removed and can be fitted to model M_b. The generalized removal method fits successively more general models until an acceptable fit is found. Then the population size is obtained based on the final selected model.

The reason for setting the average probabilities equal (except for the first few) is to reduce the number of parameters. However, it may contradict the finding of decreasing probabilities mentioned above. Under the model assumption, even if some animals with large removal probabilities are removed, heterogeneity still exists and model M_b is not theoretically valid for the remaining animals, especially when removal probabilities vary greatly. Simulation studies (Otis et al. 1978, p. 130; Pollock and

Otto 1983, p. 1043) have shown that the generalized removal estimator typically has a negative bias, which can be large if the removal probabilities are very heterogeneous. An improved estimator due to Pollock and Otto (1983) is described below.

For the first-capture data $(u_1, u_2, \ldots, u_6)$ in table 2.2, the model assuming that all removal probabilities are equal fits well. The resulting estimate of the population size is 41.0 (standard error 3.1), which is the same as obtained with model M_b. The average removal probability is estimated to be 0.34, which is identical to the estimated first-capture probability in model M_b.

MODEL M_{th}: $P_{ij} = p_i e_j$

This model combines time and heterogeneity effects, with a multiplicative form of the effects being assumed. No estimator of population size had been suggested at the time that the paper by Otis et al. (1978) was published.

In some capture–recapture studies for large animals, actual recapture is not necessary and records of resighting suffice to provide the necessary recapture information. In this case, there is no need to model any behavioral response to capture. As a consequence, model M_{th} may be a reasonable model. It can also be considered when different trapping methods are feasible. In applications to human populations, usually different ascertainment methods (e.g., occurrence on different lists of patients) are used to "capture" individuals. Hence, this model is potentially useful in analyzing multiple-list problems. Recent developments of this model are considered below.

MODEL M_{tbh}: $P_{ij} = p_{ij}$ UNTIL FIRST CAPTURE AND
$P_{ij} = c_{ij}$ FOR ANY RECAPTURE

In this model, three sources of variations in capture probabilities are considered. This most general model has previously been considered only conceptually useful and too complicated to be applied to practical situations if no restrictions are made. However, the general model M_{tbh} has been selected as the most likely model for estimating the size of squirrel (White et al. 1982, p. 149) and mouse (Otis et al. 1978, p. 93) populations. If reasonable assumptions can be made, then an estimation procedure for this model would offer a unified approach for all eight models. New methodologies that usually require extensive numerical computations have been developed and are discussed below.

New Approaches

In the past two decades, there has been an explosion of methodological research on capture–recapture. In the remainder of the chapter some of these methods are described and illustrated using the data of table 2.1.

BOOTSTRAP METHODS

For models without individual heterogeneity, standard likelihood theory yields a large-sample variance estimator for the MLE. Often this variance approximation may not be adequate for small samples. Moreover, for heterogeneous models, the corresponding likelihood functions are mathematically intractable. The use of resampling methods such as the bootstrap has been proposed (Efron and Tibshirani 1993; Buckland and Garthwaite 1991) to obtain variance estimators. The bootstrap is applicable to a variety of capture–recapture models (Chao et al. 2001b). The procedure is to resample capture–recapture data according to the relative frequencies of the observable capture patterns. Then a bootstrap estimate of population size is calculated using the resampled data. After many replications, the variance is estimated as the sample variance of all bootstrap estimates. Resampling techniques can also be used for obtaining confidence intervals.

IMPROVED INTERVAL ESTIMATION

The construction of traditional confidence intervals is based on the large-sample normality of population size estimators. Several drawbacks for this method have been noted (Chao 1987): (1) the sampling distribution is usually skewed; (2) the lower bound of the resulting interval may be less than the number of animals caught in the experiment; and (3) the coverage probabilities are often unsatisfactory. To overcome these problems, a log-transformation was suggested by Burnham and applied by Chao (1987). That is, the 95% confidence interval for any estimator $\hat{N}$ with an estimated variance $\hat{\sigma}^2$ can be constructed as

$$M_{k+1} + \frac{\hat{N} - M_{k+1}}{C} \text{ to } M_{k+1} + (\hat{N} - M_{k+1}) \times C \qquad (4.2)$$

where

$$C = \exp\left\{ 1.96 \sqrt{\ln\left[1 + \frac{\hat{\sigma}^2}{(\hat{N} - M_{k+1})^2} \right]} \right\}$$

The lower bound of this interval is always greater than the number of different animals actually captured in all occasions. Alternative approaches include the resampling methods mentioned above and the use of profile likelihoods (Cormack 1992).

MAXIMUM LIKELIHOOD ESTIMATOR FOR MODEL M_{tb}

Burnham derived the MLE of population size for model M_{tb} under the assumption that recapture probability, c_j, $j = 2, \ldots, k$, is a power function of the initial capture probability, p_j; i.e., $c_j = p_j^{1/\theta}$, for some parameter θ (Rexstad and Burnham 1991, p. 13). This MLE has been implemented in program CAPTURE. A parameter $\theta > 1$ implies that the animals exhibit trap-happy behavior, while $\theta < 1$ implies trap-shy behavior. Using CAPTURE, for the mouse data in table 2.1 $\hat{N} = 40$ (standard error 3.6), $\hat{\theta} = 2.09$, $(\hat{p}_1, \hat{p}_2, \hat{p}_3, \hat{p}_4, \hat{p}_5, \hat{p}_6) = (0.37, 0.39, 0.26, 0.28, 0.43, 0.42)$, $-2\ln(L_{tb}) = 94.6$ and AIC = 110.6.

A problem with this estimator is that the iterative steps may fail to converge even for nonsparse data. This motivated Chao et al. (2000) to find the MLE under the alternative assumption that the recapture probabilities bear a constant relationship to initial capture probabilities. That is, $c_j / p_j = \theta$ for all $j = 2, \ldots, k$. The likelihood can be formulated (Chao et al. 2000) and an approximate variance derived. The case $\theta > 1$ corresponds to trap-happy behavior and $\theta < 1$ corresponds to trap-shy behavior. Using CARE-2 (a computer program introduced in section 4.4), and fitting the model to the data in table 2.1, gives $\hat{\theta} = 2.34$, $(\hat{p}_1, \hat{p}_2, \hat{p}_3, \hat{p}_4, \hat{p}_5, \hat{p}_6) = (0.34, 0.33, 0.21, 0.23, 0.29, 0.27)$ and $\hat{N} = 44.0$ (standard error 6.9). Also, for this model, $-2\ln(L_{tb}) = 92.3$ and AIC = 108.3.

The two versions of model M_{tb} applied to the mouse example show strong trap-happy behavior for the animals and slight time-varying effects. Each model is a significantly better fit to the data than model M_t (see section 1.4) but is not significantly different from model M_b. Thus, both versions of model M_{tb} and model M_b are appropriate for use in estimating population size. Model M_b is selected among the three adequate models because the AIC for this model is the lowest.

When time and behavioral response affect capture probabilities, the two approaches above provide applicable models. However, both models need sufficient capture information to allow for stable estimation, especially when there are many capture occasions.

JACKKNIFE TECHNIQUE FOR MODEL M_{bh}

When there are numerous parameters such as models M_h and M_{bh}, a nonparametric technique such as the jackknife has been used to construct new estimators. For model M_h, there is Burnham and Overton's jackknife as given in equation 4.1. For model M_{bh}, Pollock and Otto (1983) recommended another type of jackknife, which was shown in simulations to be an improvement over the generalized removal estimator. This jackknife estimator has the simple form

$$\hat{N} = u_1 + u_2 + \cdots + u_{k-1} + ku_k = M_{k+1} + (k-1)u_k \qquad (4.3)$$

It implies that the number of missing animals is estimated by $(k-1)u_k$, depending only on the first captures caught on the last occasion. When nearly all animals are captured in the experiment so that several animals are newly caught on the last occasion, this estimator tends to overestimate. For the frequency data in table 2.2, the jackknife estimate is 53, which is much higher than the estimate from other models.

SAMPLE COVERAGE APPROACHES FOR MODELS M_h AND M_{th}

The sample coverage approach was originally applied to heterogeneous species abundance models (Chao and Lee 1992). Then Chao et al. (1992) and Lee and Chao (1994) modified it for application to capture–recapture experiments with emphasis on the models M_h and M_{th}. They assumed that the heterogeneity effects can be summarized in terms of the mean and coefficient of variation (CV) of $p_1, p_2, \ldots,$ p_N. This CV is a nonnegative parameter that characterizes the degree of heterogeneity. The CV is zero if and only if the animals are equally catchable, and the larger the CV, the greater the degree of heterogeneity among animals.

The sample coverage is defined as the proportion of the p_i's that is associated with the individual variation among captured animals and it can be estimated in heterogeneous models. When there are at least five occasions, Chao et al. (1992) and Lee and Chao (1994) suggested the sample coverage estimator

$$\hat{C}_1 = 1 - \frac{f_1}{\sum_{j=1}^k jf_j}$$

and its bias-corrected form

$$\hat{C}_2 = 1 - \frac{f_1 - 2f_2/(k-1)}{\sum_{j=1}^k jf_j}$$

Then the population size can be estimated via the estimation of sample coverage. The two population size estimators for model $\mathbf{M}_h$ are

$$\hat{N}_i = \frac{M_{k+1}}{\hat{C}_i} + \frac{f_1}{\hat{C}_i}\hat{\gamma}_i^2 \qquad i = 1, 2 \qquad (4.4)$$

where

$$\hat{\gamma}_i^2 = \max\left[\frac{kM_{k+1}}{(k-1)\hat{C}_i(\Sigma\, jf_j)^2}\sum_{j=1}^{k} j(j-1)f_j - 1, 0\right]$$

is an estimator for CV^2. For model $\mathbf{M}_{th}$, the estimators have similar form except for a different CV estimator (Lee and Chao 1994).

There are several advantages to the sample coverage approach: (1) it utilizes all the capture frequencies and has an explicit form; (2) a measure is provided to quantify the degree of heterogeneity; and (3) the estimator tends to the true population size when $p_1, p_2, \ldots, p_N$ are a random sample from a gamma distribution. The limitation is that a sufficient number of capture occasions (say, at least five) and sufficient capture information are needed to accurately estimate the CV parameter. The sample coverage approach applied to the data in table 2.1 yields, with model $\mathbf{M}_h$, the two CV estimates $\hat{\gamma}_1 = 0.50$ and $\hat{\gamma}_2 = 0.48$, which gives some evidence of heterogeneity. The two estimators in (4.4) are respectively $\hat{N}_1 = 44$ (standard error 3.8) and $\hat{N}_2 = 42$ (standard error 3.5). The results from model $\mathbf{M}_{th}$, are not presented here, but were quite similar.

ESTIMATING EQUATIONS (INCLUDING MARTINGALE METHOD AND MAXIMUM QUASI-LIKELIHOOD)

Chao et al. (2001b) explored a general estimating equation and provided a unified approach for a special case of model $\mathbf{M}_{tbh}$. The first-capture probability for the ith animal is designated $P_{ij} = p_i e_j$. This model combines the time effect for the jth occasion, e_j, and the heterogeneity effect for animal i, p_i, in a multiplicative form. Similarly, the recapture probability is $C_{ij} = \theta p_i e_j$ (i.e., the recapture probability is proportional to the first-capture probability with the same constant of proportionality for all animals). To reduce the number of parameters and remove nonidentification, it was also assumed that the heterogeneity effects $p_1, p_2, \ldots, p_N$ are characterized by their mean and CV.

Although the likelihood is not generally obtainable, the moments (i.e., the expected values and variances of u_j and m_j) of first capture and recapture, for each capture occasion j, can be derived in terms of the mean and CV of the heterogeneity effects. Consequently, an estimating equation approach is available without much further restriction. This approach extends the martingale approach (Yip 1991; Lloyd 1994) and maximum quasi-likelihood method (Chao et al. 2000). This approach is computer intensive and a large number of capture and recapture data are needed to estimate time, heterogeneity, and behavioral effects. Otherwise, the iterations for estimation may not converge. As in all modeling, the more complex the models, the more data are required to obtain reliable parameter estimates.

When the estimating equation approach is applied to the data in table 2.1, the following population size estimates are obtained, along with estimated bootstrap standard errors for the heterogeneous models: model M_h: 40.0 (standard error 2.1); model M_{th}: 40 (standard error 2.2); model M_{bh}: 44 (standard error 4.4); model M_{tbh}: 44 (standard error 5.4). Unfortunately, the usual likelihood-based tests cannot differentiate between these models because the likelihood functions are not computable. See below for further discussion on model selection and model uncertainty.

It has been seen from fitting model M_b that the data show strong evidence of the trap-happy phenomenon. Recall that model M_b produces a population size estimate of 41.0 (standard error 3.1) with a 95% confidence interval from equation 4.2 of (39 to 53). In addition to the behavioral response, the further general models M_{bh} and M_{tbh} merit consideration. For this data set these models produce close results, so it is reasonable to adopt the most general model M_{tbh} and conclude that the population size is about 44 (standard error 5.4). The data based on model M_{tbh} show strong trap-happy behavior ($\hat{\theta} = 1.89$), a low degree of heterogeneity (the CV estimate is 0.36), and slight time-varying effects as the relative time effects are estimated to be ($\bar{p}e_1, \bar{p}e_2, \ldots, \bar{p}e_k$) = (0.34, 0.32, 0.26, 0.26, 0.33, 0.33), where $\bar{p}$ denotes the average of p_i's.

Usually, a simpler model has smaller variance, but larger bias, whereas a general model, because it includes more parameters, has lower bias but larger variance. The 95% confidence interval using (4.2) under model M_{tbh}, for example, is 40 to 65. An objective selection rule between models M_b and M_{tbh} is currently not available. A simpler model produces narrow confidence interval with possibly poor coverage probability, whereas a more general model produces wide interval with more satisfactory coverage probability. This trade-off is clear in comparisons of the different models built around our deer mouse data.

OTHER APPROACHES

Some other important methods will now be briefly considered. See the relevant references for details and software.

Log-linear or Generalized Linear Models

This approach models the observable frequency counts of unique capture histories as a log-linear model or a generalized linear model. Then a model is selected and used to estimate the missing cells. Model selection and model testing then can be performed under a unified framework (Fienberg 1972; Cormack 1989; Agresti 1994; Coull and Agresti 1999; Evans et al. 1994; IWGDMF 1995). The simplest independent log-linear model is equivalent to the classical model M_t discussed in chapter 2. Chao et al. (2001a) describes the relative merits of this approach.

Bayesian Methods

In the past, a limitation of classical Bayesian estimation approaches was the complex calculations involved. Now this problem can be overcome by the use of Gibbs sampling, a Monte Carlo Markov Chain technique (Castledine 1981; Smith 1988, 1991; George and Robert 1992; Lee and Chen 1998). With the Bayesian approach, empirical information can be incorporated and nonidentifiable models can also be treated (Lee and Chen 1998). However, there have been arguments about the sensitivity of the Bayesian estimates to the choice of priors (Chao 1989). Huggins (2002) has developed an empirical Bayes approach for model M_h assuming a beta distribution of the capture probabilities.

Parametric Approaches for Modeling Heterogeneity

Lloyd and Yip (1991), Sanathanan (1972), and Coull and Agresti (1999), respectively, used beta, gamma, and normal distributions to model heterogeneity. This approach reduces the number of parameters to only a few and thus traditional inference can be easily applied. A difficulty in this approach is that several models may adequately fit the data, but the estimated population sizes may differ greatly among models (Link 2003). It is important to remember that many difficulties can result from assuming a particular mathematical form for data without including an appropriate level of biological intuition to check whether the assumptions are compatible with common sense.

Latent Class, Mixture Model, and Nonparametric Maximum Likelihood

This approach assumes that the population can be partitioned into two or more homogeneous groups (Darroch et al. 1993; Agresti 1994; Norris and Pollock 1996b; Coull and Agresti 1999; Pledger 2000) so that a latent class or a finite mixture can be applied to model capture probability. Pledger (2000) found that for many data sets a simple dichotomy of animals is enough to correct for bias due to heterogeneity. See Link (2003) for relevant discussion.

Model Selection, Model Uncertainty, and Model Averaging

Model selection is an important topic in statistical inference (Buckland et al. 1997). Currently preferred model selection procedures using likelihood ratio tests or the AIC criterion (among models for which likelihoods are obtainable) were discussed in section 1.4. Stanley and Burnham (1998) commented that the existing model selection procedure in CAPTURE (Otis et al. 1978) usually selects an inappropriate model in simulations. For models with heterogeneity, the likelihoods are usually not computable unless parametric assumptions are made. Currently there is no objective method to select a model from the various heterogeneous models. Stanley and Burnham (1998) proposed the use of a model-averaging concept, in which weighted estimates from competing models are combined. Norris and Pollock (1996a) also considered model uncertainty in the estimation of variances.

Table 4.1 (Chao 2001) summarizes for each model the available approaches that are known. References are provided in the table for modeling approaches that have not been addressed in the above discussion.

We remark that capture–recapture models have wide applications to nonanimal populations in other disciplines. Some applications include the estimation of plant populations (Alexander et al. 1997), elusive human populations (e.g., cases for a specific diseases, homeless persons, and mentally ill people [IWDGMF 1995]), the size of vocabulary for an author in linguistics (Efron and Thisted 1976), the number of undetected bugs in a piece of software in software reliability (Briand et al. 2000), and the number of undiscovered genes in genetics (Huang and Weir 2001).

Models Incorporating Covariates

Individual, environmental, or other auxiliary or concomitant variables recorded or measured in capture–recapture experiments commonly are called covariates. Using such variables, sophisticated models may be fitted that both estimate the population size and give insight into the capture process (Huggins 1989, 1991; Alho 1990).

Pollock et al. (1984) proposed a model to incorporate covariates in the analysis. They developed an estimation procedure based on the full likelihood. However, the covariates for the uncaptured animals are not observable. Therefore, they had to categorize all animals into several groups and use the midpoint of the classifying covariate as a representative

TABLE 4.1
Discrete-time models and associated estimation methods

Model	Approaches/methods
M_{tbh}	Estimating equation[a] (Chao et al. 2001b) Generalized linear models using covariates[a,b] (Huggins 1989, 1991) Conditional sample coverage (Lee 1996) MLE using mixture (Pledger 2000) (All approaches are valid for the submodels given below; some of them are not repeated in each model)
M_{bh}	Generalized removal[c] (Otis et al. 1978) Jackknife[a,c] (Pollock and Otto 1983) Sample coverage[a] (Lee and Chao 1994) Estimating equation[a] (Chao et al. 2001b) Nonparametric MLE (Norris and Pollock 1995, 1996b)
M_{tb}	MLE[a,b,c] (Rexstad and Burnham 1991; Chao et al. 2000) Estimating equation or martingale method[a] (Lloyd 1994; Chao et al. 2001b) Bayes approach (Lee and Chen 1998) Log-linear models (Cormack 1989)
M_{th}	Sample coverage[a,c] (Lee and Chao 1994) Estimating equation[a] (Chao et al. 2001b) Mixed logit models (Agresti 1994; Coull and Agresti 1999) Log-linear models or latent class models (Darroch et al. 1993; Agresti 1994)
M_h	Jackknife method[a,c] (Burnham and Overton 1978) Moment estimator[c] (Chao 1987) Sample coverage[a] (Lee and Chao 1994) Estimating equation[a] (Chao et al. 2001b) Martingale method (Lloyd and Yip 1991) Bootstrap estimator (Smith and van Belle 1984) Empirical Bayes approach (Huggins 2002) Nonparametric MLE (Norris and Pollock 1995, 1996b)
M_b	MLE[a,b,c] (Zippin 1956) Estimating equation or martingale method[a] (Lloyd 1994) Log-linear models (Cormack 1989)
M_t	MLE[a,b,c] (Darroch 1958; Otis et al. 1978) Moment estimator[c] (Chao 1989) Estimating equation or martingale method[a] (Yip 1991) Binomial model (Schnabel 1938) Log-linear models (Fienberg 1972) Bayes and empirical Bayes (Castledine 1981; Smith 1988, 1991; George and Robert 1992)
M_0	MLE[a,b,c] (Otis et al. 1978)
	(All estimators given above are valid here)

[a] Available in software CARE-2.   [b] Available in software MARK.   [c] Available in software CAPTURE.
Source. Chao (2001).

covariate for a group. Huggins (1989, 1991) avoided this difficulty by using a likelihood conditional on the captured animals, so that the co-variates of uncaptured animals are not needed in the analysis.

To illustrate the covariate analysis, note that there were actually three covariates—sex (male or female), age (young, semi-adult or adult), and weight—collected for each individual in the deer mouse data of table 2.1. The complete data are shown in table 4.2. Only three semi-adult mice were caught, so they were reclassified as adults. The variables sex and age are categorical and each has two categories. To include these covari-ates in the model, define the following dummy variables:

W_{i1} = 1 if the ith animal is a male, and 0 for a female; and
W_{i2} = 1 if the ith animal is a young, and 0 for an adult.

The covariate weight is a continuous variable, thus

W_{i3} = weight of the ith animal.

Huggins (1989, 1991) proposed a logistic type of model for the first-capture probability of the ith animal on the jth occasion, P_{ij}, and also a similar form for the recapture probability C_{ij} as follows:

$$P_{ij} = \frac{\exp(a + a_j + \beta_1 W_{i1} + \beta_2 W_{i2} + \beta_3 W_{i3})}{1 + \exp(a + a_j + \beta_1 W_{i1} + \beta_2 W_{i2} + \beta_3 W_{i3})} \tag{4.5a}$$

$$C_{ij} = \frac{\exp(a + a_j + v + \beta_1 W_{i1} + \beta_2 W_{i2} + \beta_3 W_{i3})}{1 + \exp(a + a_j + v + \beta_1 W_{i1} + \beta_2 W_{i2} + \beta_3 W_{i3})} \tag{4.5b}$$

Here the parameter a denotes the intercept, a_j denotes the time-effect (with $a_k = 0$ because the last occasion is used as a basis for compari-son), β_1 denotes the effect of being male, β_2 denotes the effect of being young, β_3 denotes the effect of a unit of weight, and v denotes the ef-fect of behavioral response. The above model is denoted by M_{tbh}^* to distinguish it from the model without using covariates. Under this model, the individual heterogeneity is assumed to be fully determined by the observable covariates. The parameters β_1, β_2, β_3, a_1, a_2, . . . , a_{k-1}, and v denote, respectively, the heterogeneity, time effects, and be-havioral response effect.

The interpretation of the coefficient of any β is based on the fact that when $\beta > 0$, the larger the covariate is, the larger the capture probability is, while if $\beta < 0$ then the larger the covariate is, the smaller the capture probability is. Also, the parameter v represents the effect of a recapture, which implies that $v > 0$ corresponds a case of trap happy and $v < 0$ cor-responds to the case of trap shy.

TABLE 4.2
Individual capture history of mouse with three covariates: sex (m: male, f: female), age (y: young, sa: semi-adult, a: adult), and weight (in grams)

Occasion 1	Occasion 2	Occasion 3	Occasion 4	Occasion 5	Occasion 6	Sex	Age	Weight
1	1	1	1	1	1	m	y	12
1	0	0	1	1	1	f	y	15
1	1	0	0	1	1	m	y	15
1	1	0	1	1	1	m	y	15
1	1	1	1	1	1	m	y	13
1	1	0	1	1	1	m	a	21
1	1	1	1	1	0	m	y	11
1	1	1	0	0	1	m	sa	15
1	1	1	1	1	1	m	y	14
1	1	0	1	1	1	m	y	13
1	1	0	1	1	1	m	y	14
1	1	1	0	1	1	f	a	22
1	1	1	1	1	1	m	y	14
1	0	1	1	1	0	m	y	11
1	0	0	1	0	0	f	y	10
0	1	0	0	1	0	f	a	23
0	1	1	0	0	1	f	y	7
0	1	0	0	0	1	m	y	8
0	1	0	1	0	1	m	a	19
0	1	1	0	1	0	m	y	13
0	1	0	1	0	1	f	y	5
0	1	0	0	0	1	f	a	20
0	1	0	0	1	1	m	y	12
0	0	1	0	0	0	f	y	6
0	0	1	1	1	1	f	a	22
0	0	1	0	1	1	f	y	10
0	0	1	1	1	1	f	y	14
0	0	1	0	1	0	f	a	19
0	0	1	0	0	0	f	a	19
0	0	0	1	0	0	f	a	20
0	0	0	1	1	1	m	sa	16
0	0	0	1	1	0	f	y	11
0	0	0	0	1	0	m	y	14
0	0	0	0	1	0	f	y	11
0	0	0	0	1	0	m	a	24
0	0	0	0	0	1	m	y	9
0	0	0	0	0	1	m	sa	16
0	0	0	0	0	1	f	a	19

It is sometimes more convenient to rewrite (4.5a) and (4.5b) as logit functions:

$$\text{logit}(P_{ij}) = \ln\left(\frac{P_{ij}}{1 - P_{ij}}\right) = a + a_j + \beta_1 W_{i1} + \beta_2 W_{i2} + \beta_3 W_{i3}$$

for the first-capture probability, and

$$\text{logit}(C_{ij}) = \ln\left(\frac{C_{ij}}{1 - C_{ij}}\right) = a + a_j + v + \beta_1 W_{i1} + \beta_2 W_{i2} + \beta_3 W_{i3}$$

for the recapture probability.

To construct the "conditional" likelihood proposed in Huggins (1989, 1991), consider, for example, the first mouse captured. Table 4.2 shows that this was a young male with weight 12, so that $W_{11} = 1$, $W_{12} = 1$, and $W_{13} = 12$. With these covariates in equation 4.5, the probability for the first animal with capture history 111111 is $P_{11}C_{12}C_{13}C_{14}C_{15}C_{16}$. The idea here is that given the event that this mouse was captured, with probability

$$1 - \prod_{j=1}^{k}(1 - P_{1j})$$

the likelihood is expressed as

$$\frac{P_{11}C_{12}C_{13}C_{14}C_{15}C_{16}}{1 - \prod_{j=1}^{k}(1 - P_{1j})}$$

Similarly, for the second young female mouse with weight 15 and history (100111), $W_{21} = 0$, $W_{22} = 1$, and $W_{23} = 15$ and the conditional likelihood is

$$\frac{P_{21}(1 - C_{22})(1 - C_{23})C_{24}C_{25}C_{26}}{1 - \prod_{j=1}^{k}(1 - P_{2j})}$$

The full conditional likelihood is a product of 38 individual likelihoods and is a function of a, a_1, a_2, ..., a_5, β_1, β_2, β_3, and v. Therefore, numerical methods must be used for finding the (conditional) MLEs. After the MLEs of the parameters are obtained, the population size is estimated by the Horvitz-Thompson (1952) estimator,

$$\hat{N}_{HT} = \sum_{i=1}^{M_{k+1}}\left[1 - \prod_{j=1}^{k}(1 - \hat{P}_{ij})\right]^{-1}$$

TABLE 4.3
Estimation results for various models with covariates

	M_{tbh}^*	M_{bh}^*	M_{tb}^*	M_{th}^*	M_h^*	M_b^*	M_t^*	M_0^*
$\hat{N}$	47.1	47.2	46.5	39.7	39.9	42.3	38.4	38.5
SE	10.4	7.2	12.7	1.6	1.7	3.8	0.7	0.7
$-2\ln(L)$	274.6	279.1	296.4	279.1	289.7	300.8	304.8	314.5
Parameters	10	5	7	9	4	2	6	1
AIC	294.7	289.1	310.4	297.1	297.7	304.9	316.8	316.5

where $\hat{P}_{ij}$ is the estimated capture probability evaluated at the conditional MLE. The variance of the resulting estimator can be estimated by an approximate variance formula derived in Huggins (1989, 1991).

The logit approach allows the use of standard linear models to model the capture probabilities and conditional versions of the likelihood ratio test and AIC to compare submodels. Submodels of the most general model M_{tbh}^* corresponding to those of Otis et al. (1978) are

M_{bh}^* : set $a_1 = a_2 = \cdots = a_5 = 0$ in model M_{tbh}^*;
M_{tb}^* : set $\beta_1 = \beta_2 = \beta_3 = 0$ in model M_{tbh}^*;
M_{th}^* : set $v = 0$ in model M_{tbh}^*;
M_h^* : set $a_1 = a_2 = \cdots = a_5 = 0$, $v = 0$ in model M_{tbh}^*;
M_b^* : set $a_1 = a_2 = \cdots = a_5 = 0$, $\beta_1 = \beta_2 = \beta_3 = 0$ in model M_{tbh}^*;
M_t^* : set $v = 0$, $\beta_1 = \beta_2 = \beta_3 = 0$ in model M_{tbh}^*; and
M_0^* : set $a_1 = a_2 = \cdots = a_5 = 0$, $\beta_1 = \beta_2 = \beta_3 = 0$, $v = 0$ in model M_{tbh}^*.

Table 4.3 shows the results for fitting these models to the deer mouse data, using the software CARE-2 (to be introduced in section 4.4).

For each model, the corresponding estimated population size and its standard error, twice the maximized log-likelihood, the number of parameters and the Akaike Information Criterion (AIC) are shown. From the values of AIC in table 4.3, model M_{bh}^* is selected for this data set. Based on the selected model, $\hat{\beta}_1 = 0.918$ is the effect for a male (the female is set to be 0), so males have larger capture probabilities than females, and $\hat{\beta}_2 = 1.887$ is the effect for a young (the adult effect is set to be 0), so young have larger capture probabilities than adults. The coefficient $\hat{\beta}_3 = 0.16$ is the effect for a unit change of body weight. This implies the capture probability increases with the weight. Also $\hat{v} = 1.179$, which provides evidence of trap-happy behavior. The estimated population size under the selected model M_{bh}^* is 47 (standard error 7.2), and a 95% confidence interval using (4.2) is 40 to 74. This logistic model simultaneously models a behavioral response and heterogeneity based on the observable covariates. Recall that if covariates are ignored, the estimating equation approach for both models M_{bh} and M_{tbh} yield an estimate of 44, and the latter provides a 95% confidence interval of 40 to 65.

Advantages for models incorporating covariates include the following: (1) the models provide a clear description of the sources of heterogeneity, and the effect of each covariate can be assessed; (2) model comparisons and model selection can be performed under a unified framework; and (3) if all relevant covariates are included, then these models generally yield better estimators with respect to bias and precision, though the above analysis of the mouse data could not show this. One assumption is that the heterogeneity effects are fully determined by the observable covariates. The use of covariates does not take into account the heterogeneity due to unobservable innate characteristics of the individuals. If some important covariates were not recorded, the models might work less well than those without using covariates.

4.3 Continuous-time Models

For insects or large mammals, trapping is often conducted over a fixed period of time, and captures (or sightings) of individuals can occur at any time during the data collection. Thus, there is only one capture on each capture occasion (i.e., sample size of 1) and the exact time for each capture is recorded. This type of experiment is referred to as a continuous-time capture–recapture experiment.

Data Structure

An early paper by Craig (1953) described a continuous-time experiment, in which one butterfly was netted at a time, marked with a spot of nail polish, and then released immediately. Darroch (1958) discussed the analysis for samples of size 1, though these authors did not refer to their methods as continuous-time models. In most large-mammal studies, usually only one individual can be captured or sighted at a time. Actual capture or marking is not always necessary because large animals can sometimes be uniquely identified by physical characteristics, such as shape, scars, color variation in fur, geographically separated territories, etc. This type of continuous-time model is therefore particularly useful for studies based on sightings or photographs of animals such as whales (Whitehead and Arnbom 1987) and bears (Mace et al. 1994; Boyce et al. 2001; Keating et al. 2002). Some of the bear-sighting data sets of Boyce et al. (2001) and Keating et al. (2002) are used for an example.

TABLE 4.4
Hypothetical example of continuous-time data

Animal	Capture times	Capture frequency
1	1.6, 4.0, 5.3	3
2	2.5	1
3	3.7, 4.3, 5.1, 6.6	4
4	6.3	1
5	(No capture)	0
6	(No capture)	0

The model being considered is also a general framework for species richness estimation where a "unique individual" is equivalent to a species and a "capture" means that a species is represented in a sample. A pioneering paper in this area is by Fisher et al. (1943). A review is provided by Bunge and Fitzpatrick (1993), and Chao (2005) and Wilson and Anderson (1995) and Wilson and Collins (1993) provided some simulation comparisons of estimation methods.

For each captured animal the complete history consists of a series of exact capture times. A small hypothetical example will be used to explain the data structure. Assume that there are 6 animals in a population $(N = 6)$ and four animals are caught in a total of 9 captures in the time period of $(0, T)$. The capture times for the four animals and the capture frequencies shown in table 4.4.

For the models to be discussed below, the statistics needed for estimation are the frequency counts $f_1, f_2, \ldots, f_s$, or functions of these counts, where s denotes the maximum frequency and f_j denotes the number of individuals that are captured exactly j times in the time interval $(0, T)$. The complete capture information for the above data can therefore be summarized as follows, where M denotes the number of distinct animals caught and n the total number of captures:

$$f_1 = 2, \ f_2 = 0, \ f_3 = 1, \ f_4 = 1, \ M = 4, \ \text{and} \ n = 9$$

Out of 9 captures, 4 distinct individuals were captured; two of them were captured once, one was captured three times, and one was captured four times. It is clear that $f_1 + f_2 + \cdots + f_s = M$ and $f_1 + 2f_2 + \cdots + sf_s = n$.

The Classical Homogenous Model

Assume that the capture or sighting process is a Poisson process with rate parameter λ in the time period of $(0, T)$. This means that the probability of capturing an animal in a small interval around t, say $(t, t + dt)$, is approximately $\lambda\, dt$, and the captures in two nonoverlapping intervals are independent. The number of captures in any time interval (t_1, t_2) is a Poisson random variable with parameter $\lambda(t_2 - t_1)$, which implies that the probability of no capture in this interval is $\exp[-\lambda(t_2 - t_1)]$.

With the classical homogeneous model the Poisson rate parameters for all animals are identical and constant over time. This model is a continuous-time version of model M_0. The intuitive approach of Crowder et al. (1991, p. 165) can be used to obtain the likelihood. For the first captured animal in the example, consider the following consecutive events in nonoverlapping intervals: no capture in $(0, 1.6)$, one capture in $(1.6, 1.6 + dt_1)$, no capture in $(1.6 + dt_1, 4.0)$, one capture in $(4.0, 4.0 + dt_2)$, no capture in $(4.0 + dt_2, 5.3)$, one capture in $(5.3, 5.3 + dt_3)$, and no capture in $(5.3 + dt_3, T)$. Then the joint probability of these events is equal to the product of the probabilities as given by

$$\exp(-1.6\lambda)\, \lambda\, dt_1 \exp[-\lambda(4.0 - 1.6 - dt_1)]\, \lambda\, dt_2$$
$$\times \exp[-\lambda(5.3 - 4.0 - dt_2)]\, \lambda\, dt_3 \exp[-\lambda(T - 5.3 - dt_3)]$$

Dividing the above product probability by $dt_1 dt_2 dt_3$ and letting dt_1, dt_2, dt_3 all tend to zero gives the likelihood for the first animal as $\lambda^3 \exp(-T\lambda)$. It is clear that only the capture frequency is relevant in the estimation procedure. Similarly, the likelihoods for the second and the third animals are, respectively, $\lambda \exp(-T\lambda)$ and $\lambda^2 \exp(-T\lambda)$. For the animals that are not captured, the probability is $\exp(-T\lambda)$. Moreover, in a homogeneous model, the number of possible divisions of 6 individuals into 5 frequency groups with frequencies 0, 1, 2, 3, 4 and sizes 2, 2, 0, 1, 1 is $6!/(2!\, 2!\, 0!\, 1!\, 1!)$. As a result, the full likelihood for the data is

$$L = \frac{6!}{2!\, 2!\, 0!\, 1!\, 1!}\, \lambda^9 \exp(-6T\lambda)$$

Generally, under a homogeneous model, the full likelihood becomes

$$L = \frac{N!}{(N - M)!\, f_1!\, f_2! \cdots f_s!}\, \lambda^n \exp(-N\lambda T)$$

In this special homogenous case, the statistics M and n are sufficient for estimating the parameters N and λ and the MLE can thus be obtained by maximizing the above likelihood. It follows from Craig (1953, p. 172) that the exact MLE of N is the solution of the following equation:

$$\sum_{j=1}^{M} (N - j + 1)^{-1} = \frac{n}{N}$$

Darroch (1958, p. 349) obtained an approximate MLE as the solution of the equation

$$1 - \frac{M}{N} = \left(1 - \frac{1}{N}\right)^n$$

An approximate variance estimator can be obtained by substituting an estimate of N into the variance formula

$$\text{Var}(\hat{N}) = \frac{N}{\exp(n/N) - 1 - n/N}$$

Becker (1984) and Becker and Heyde (1990) extended the above model to the case where the Poisson rate is a time-varying function, $\lambda(t)$ in the interval $(0, T)$, i.e., for the continuous-time version of model M_t. These authors found that the estimation procedures are identical to those for model M_0.

Heterogeneous Models Using Bear-resighting Data

The homogeneity assumption is questionable in most applications. A heterogeneous model M_h can therefore be considered. In this model, capture rates are allowed to vary among individuals by assuming that the capture process for the ith animal is a Poisson process with rate λ_i. As in discrete-time models, assumptions are needed to model the parameters $\lambda_1, \lambda_2, \ldots, \lambda_N$. Parametric approaches have been adopted by assuming that $\lambda_1, \lambda_2, \ldots, \lambda_N$ are selected from a distribution. A common choice is the gamma distribution, which was proposed by Greenwood and Yule (1920) and Fisher et al. (1943). Tanton (1965) and many subsequent authors have applied it to wildlife studies, including recent applications to bears (Boyce et al. 2001; Keating et al. 2002). The model is referred to as a Poisson-gamma model or a negative binomial model because the capture frequencies follow a negative binomial distribution. See Chao and Bunge (2002) for details of the likelihood and the estimation procedures. Other possible parametric family includes the log-normal (Bulmer 1974) and the inverse Gaussian (Ord and Whitmore 1986).

The continuous-time model can be regarded as a special case of the discrete-time model with all sample sizes being one. The jackknife and sample coverage approaches discussed in discrete-time model M_h can be applied directly here by replacing M_{k+1} and k by M and n. That is, the first-order and the second-order jackknife estimators in equation 4.1 now become

TABLE 4.5
Sighting frequencies of female grizzly bears

Year	f_1	f_2	f_3	f_4	f_5	f_6	f_7	M	n
1996	15	10	2	1	0	0	0	28	45
1997	13	7	4	1	3	0	1	29	65
1998	11	13	5	1	1	0	2	33	75

$$\hat{N}_{j1} = M + \left[\frac{n-1}{n}\right]f_1 \approx M + f_1 \tag{4.6a}$$

$$\hat{N}_{j2} = M + \left[\frac{2n-3}{n}\right]f_1 - \left[\frac{(n-2)^2}{n(n-1)}\right]f_2 \approx M + 2f_1 - f_2 \tag{4.6b}$$

From equation 4.4, a similar sample coverage approach (Chao and Lee 1993) results in the following estimators:

$$\hat{N}_i = \frac{M}{\hat{C}_i} + \frac{f_1}{\hat{C}_i}\hat{\gamma}_i^2 \qquad i = 1, 2 \tag{4.7}$$

where the two coverage estimators are the same as before, but the estimator of the CV of $\lambda_1, \lambda_2, \ldots,$ and λ_N is slightly different. Here there is the following estimator for CV^2:

$$\hat{\gamma}_i^2 = \max\left[\frac{M}{\hat{C}_i}\sum_{j=1}^{s}\frac{j(j-1)f_j}{n(n-1)} - 1, 0\right]$$

If the sample size n is large, the two sample coverage estimates ($\hat{C}_1$ and $\hat{C}_2$), and thus the two population size estimates ($\hat{N}_1$ and $\hat{N}_2$), differ little in continuous-time models. In species number estimation some species may be very abundant, which leads to a large value for the CV. In that case, a bias-corrected estimator of the CV is needed and given in Chao and Lee (1992).

Boyce et al. (2001) and Keating et al. (2002) recorded the sighting frequencies of female grizzly bears with cubs-of-the-year in the Yellowstone ecosystem. The data for 1996–98 are used here for an example (table 4.5).

Boyce et al. (2001) considered the joint estimation of population sizes over years and assumed that the shape parameter of the gamma distribution was constant over time. Their estimates are shown in table 4.6. Keating et al. (2002) adopted a sample coverage approach and the estimate $\hat{N}_1$ based on equation 4.7 is shown for each year. The table also shows

TABLE 4.6
Estimates (with standard errors) for bear-sighting data

Model/estimator	1996	1997	1998
	Homogeneous model M_0		
MLE	42.2 (7.1)	33.3 (2.9)	37.8 (3.0)
	Heterogeneous model M_h		
Poisson-Gamma	(not obtainable)	43.6 (4.5)	42.7 (7.6)
Boyce et al. (2001)	60.0 (14.0)	43.8 (7.6)	49.4 (7.5)
Jackknife $\hat{N}_{j1}$	42.7 (5.4)	41.8 (5.0)	43.9 (4.6)
Jackknife $\hat{N}_{j2}$	47.7 (9.2)	47.7 (8.6)	42.1 (8.0)
Sample coverage $\hat{N}_1$	42.0 (7.1)	41.5 (7.6)	41.2 (5.2)

the estimates under a homogenous model (i.e., model M_0), two jackknife estimates, and the estimate obtained from a Poisson-gamma model. A disadvantage for the Poisson-gamma approach is that in some cases (e.g., for 1996) the iterative steps for MLE do not converge.

For the data of 1996, the estimated CV is $\hat{\gamma}_1 = 0$. This indicates equal sightability and it is reasonable to assume a homogeneous model. The sample coverage approach yields a very close result to that for a homogenous model. For the years of 1997 and 1998, the CV estimates are, respectively, $\hat{\gamma}_1 = 0.57$ and 0.45, which provide evidence of heterogeneity. Thus, the results under a homogeneous model would underestimate the true population size. The estimates $\hat{N}_2$ are very close to those for $\hat{N}_1$, so they are omitted in table 4.6. The first-order jackknife and the sample coverage approach show that the numbers of female bears was stable over the three years.

General Models and Assumptions

As in the discrete-time models, a series of eight continuous-time models can be postulated depending on the sources of variability in the Poisson rates due to time, behavioral response, and heterogeneity. Consider a Poisson process with parameter $\lambda_i^*(t)$, which can be intuitively interpreted as capturing animal i in a small time interval around time t. The Poisson rate for model M_{tbh} is given by

$$\lambda_i^*(t) = \begin{cases} \lambda_i \, \alpha(t) & \text{until first capture} \\ \phi\lambda_i \, \alpha(t) & \text{for any recapture} \end{cases}$$

Here $\alpha(t)$, $\{\lambda_1, \lambda_2, \ldots, \lambda_N\}$, and ϕ represent the effects of time, heterogeneity, and behavioral response, and $\alpha(t)$ is any arbitrary time-varying
function in $(0, T)$. All submodels can be formulated as shown below.
Compared with discrete-time models, there has been relatively little published research for the continuous-time counterparts. Relevant references
are given for each model:

M_{bh} : set $\alpha(t) = 1$ in model M_{tbh}

M_{tb} : set $\lambda_i = 1$ in model M_{tbh} (Yip et al. 2000; Hwang et al. 2002);

M_{th} : set $\phi = 1$ in model M_{tbh} (Becker 1984; Chao and Lee 1993;
       Yip et al. 1996);

M_h : set $\alpha(t) = 1$ and $\phi = 1$ in model M_{tbh} (Chao and Lee 1993; Yip
       and Chao 1996);

M_b : set $\alpha(t) = 1$, $\lambda_i = \lambda$ in model M_{tbh} (Yip et al. 2000);

M_t : set $\lambda_i = 1$ and $\phi = 1$ in model M_{tbh} (Becker 1984; Becker and
       Heyde 1990; Yip et al. 1993); and

M_0 : set $\alpha(t) = 1$, $\lambda_i = \lambda$ and $\phi = 1$ in model M_{tbh} (Craig 1953;
       Darroch 1958).

Beyond the classical MLE approach, Becker (1984) was the first to establish
a counting process framework for model M_t. Most subsequent authors
dealing with the other models have followed this approach. Until now, there
have been no estimation methods proposed in the literature for the most
general model M_{tbh} and model M_{bh}. However, if the heterogeneity effects
can be fully explained by observable covariates, Lin and Yip (1999) and
Hwang and Chao (2002) explored a unified likelihood-based approach.

The data collected in continuous-time format provide more information
than those in discrete-time models because the latter ignores recaptures of
the same animals in any fixed occasion. In practice, more effort is usually
needed to record the capture time for each captured animal. If times are
not recorded, that information is lost to future analysis attempts. There
has been little published work for comparing the relative merit of
continuous-time models and their corresponding discrete-time models. We
have found that estimates derived from continuous-time experiments will
generally be more precise than those from discrete-time experiments, all
other considerations being equal. It is not difficult to show that under a
classical homogeneous model that the asymptotic variance of the MLE for
population size under a continuous-time model is less than that under its
corresponding discrete-time model. A similar conclusion is expected for
other models. The gain in precision usually is most significant when only a
small fraction of animals is captured in the experiment. Then, continuous-
time models are clearly preferable to their corresponding discrete-time
models. In contrast, if a relatively large number of animals is captured,
estimators under the two types of models perform similarly (Wilson and

Anderson 1995). In this case, there is almost no difference between the two types of models.

Relationship to Recurrent Event Analysis

In many studies, individuals may experience repeated events and times of each event are recorded. For example, in industrial reliability, repeated events include failures or repairs of a system (Lawless and Thiagarajah 1996). In medical studies, repeated events include recurrences of tumors, infections, and other diseases. There is a large literature on models for the recurrent event analysis (Wei and Glidden 1997). Most of the analyses have adopted Cox's proportional hazard models based on a counting process approach (Andersen et al. 1993).

If each capture is regarded as a recurrent event then the continuous-time model can be treated as a special case of recurrent event analysis. Using this concept, the models presented above can be connected to recurrent event analysis. Many models using successive or cumulative capture times in recurrent event analysis can then be applied to capture–recapture models. However, there is a difference between the two approaches. In a recurrent event analysis, the number of individuals is usually known and the main parameter of interest is the effect of each covariate, whereas in a capture–recapture study the population size is the parameter of main interest. In a sense, these modeling approaches could be viewed as flip sides of each other. Clearly, the interaction of the two topics is a fruitful area for future research.

4.4 Computing Considerations

Information on the specialized computer programs needed for estimation for closed populations is provided in chapter 10 of this book. An additional program is necessary to accommodate recently developed estimators and the covariate methods initiated by Huggins (1989, 1991). A program CARE (for CApture-REcapture) containing three parts (CARE-1, CARE-2, and CARE-3) has been developed and is available at the website chao.stat .nthu.edu.tw. CARE-1 is an interactive S-PLUS program for the analysis of multiple lists (Chao et al. 2001a). CARE-2, a Windows 98 (or later) executable program written in the C language, calculates various estimates for the models with and without covariates that are described in section 4.2, as shown in table 4.1. The third part of CARE, CARE-3, written in GAUSS language, is an integrated program for analyzing the class of continuous-time models presented in section 4.3. A brief user guide with examples is also provided at chao.stat.nthu.edu.tw.

Figure 4.2. Flipper-tagged female Californian sea lion (*Zalophus californianus*) with her pup at Piedras Blancas, California in 1995. (Photo by Steven C. Amstrup.)

Since some special continuous-time models presented in Section 4.3 are equivalent to species estimation problems, a relevant computing program for the latter is EstimateS (Colwell 1997), which calculates various estimates of species richness including the jackknife and the sample coverage approach. The program is available from the website viceroy.eeb .uconn.edu/estimates.

4.5 Chapter Summary

- Discrete-time capture models are described. Discrete-time models assume that several trapping occasions have occurred, and that a group of animals is caught on each occasion such that the capture time of individuals within the occasion is not known.
- The eight models of Otis et al. are described as a basis for estimating population sizes, where these allow capture probabilities to vary with one or more of the factors of time, a behavioral response to capture, and individual variation in capture probabilities. Methods of estimation are discussed for each of the models.
- Newer approaches to estimation are discussed involving bootstrapping, improved confidence interval methods, parametric maximum

likelihood estimation, jackknifing, sample coverage methods, estimating equation methods, generalized linear models, Bayesian methods, parametric modeling of heterogeneity distributions, latent class models, mixture models, and nonparametric maximum likelihood.

- Model selection and related problems are discussed briefly.
- The use of models involving covariates is reviewed, including those that assume logistic regression types of functions. It is shown how this approach can be used with each of the eight models of Otis et al.
- Continuous-time models are described for situations where the capture time is known for each captured animal. For this situation the classical homogenous model is described, where captures follow a Poisson process with a constant rate parameter. Heterogeneous models (parametric and nonparametric) are then considered, where the probability of capture varies from animal to animal.
- The relationship of continuous-time models to recurrent event analysis is noted.
- Computer programs for estimation are briefly discussed.

Five

Modern Open-population Capture–Recapture Models

JAMES D. NICHOLS

5.1 Introduction

Capture–recapture studies of open populations involve multiple sample occasions at which newly captured animals are individually marked and identities of previously captured animals are recorded. Sample occasions are separated by time intervals that are sufficiently large that the population is expected to change between occasions. Hence, the population is said to be open to gains resulting from in situ reproduction and immigration and to losses from death and emigration.

Two classes of models have been developed to estimate quantities of interest from the data resulting from studies of open populations. The first class can be referred to as conditional and is exemplified by the original model of Cormack (1964; chapter 3 in this volume). This modeling is conditional on the number of animals released at each sampling occasion and permits estimation of subsequent survival and capture probabilities. The second class is unconditional and is exemplified by the models of Jolly (1965) and Seber (1965). This approach models not only the history of marked animals following release, but also the initial captures of unmarked animals, permitting estimation of such quantities as abundance, recruitment, and rate of population change.

In addition to these two approaches for modeling data from open populations, Pollock (1982) introduced a robust design that involves sampling at two temporal scales. Primary sampling periods are separated by relatively long periods, and the modeling of capture histories over these periods is based on models for open populations. However, within each primary sampling period, there are multiple secondary sampling periods separated by relatively short time intervals. The capture histories over these secondary periods are typically modeled using models for closed populations (Otis et al. 1978).

Figure 5.1. A leg-banded female Canada goose (*Branta canadensis*) defends her nest near Anchorage, Alaska. Checking nests to resight marked individuals, which are usually faithful to nesting areas, allows the estimation of the annual survival rate. (Photo by Jerry Hupp)

5.2 Conditional Single-age Models

Whereas chapter 3 provided closed-form estimators that can be computed as functions of summary statistics, this chapter focuses on models that are generally implemented using iterative techniques based on individual capture histories. A capture history (chapter 1) is simply a list of 1's and 0's, where 1 indicates that the animal was caught or observed, and 0 indicates that the animal in question was not caught or observed. For example, the capture history 01010 indicates an animal that was caught on the second and fourth sampling occasions of a 5-occasion study but not on occasions 1, 3, or 5. Every animal caught at least once during a study has a known capture history, and these data can be summarized as the number of animals exhibiting each history. For example, x_{01010} indicates the number of animals in the study exhibiting the history 01010.

Capture–recapture estimation is based on probabilistic models for events that give rise to a capture history. Conditional Cormack-Jolly-Seber (CJS) (Cormack 1964; Jolly 1965; Seber 1965) modeling of capture–recapture data for open populations is based on two primary parameters:

> p_j = the probability that a marked animal in the study population at sampling period j is captured or observed during period j; and
> ϕ_j = the probability that a marked animal in the study population at sampling period j survives until period $j + 1$ and remains in the population (i.e., the animal does not die or permanently emigrate).

In addition,

> χ_j = the probability that an animal alive and in the study population at sampling period j is not caught or observed again at any sampling period after period j.

For a capture–recapture study with k sampling periods $\chi_k = 1$, and values for sampling period $j < k$ can be computed recursively as

$$\chi_j = (1 - \phi_j) + \phi_j(1 - p_{j+1})\chi_{j+1}$$

Thus, χ_j is not really a new parameter but simply a function of the two primary parameters.

As explained in chapter 1, the modeling process can be illustrated by writing the probability corresponding to the example capture history 01010 as

$$\Pr(01010 \mid \text{release in period 2}) = \phi_2(1 - p_3)\phi_3 p_4 \chi_4$$

As the modeling approach conditions on the initial release, there are no parameters associated with capture in periods 1 or 2. However, the animal survived from period 2 to 3, and the probability associated with this event is ϕ_2. It was not captured in period 3, and the corresponding probability is given by $(1 - p_3)$; it survived from 3 to 4 (ϕ_3) and was caught (p_4), and finally it was not caught again after 4 ($\chi_4 = 1 - \phi_4 p_5$).

Note that the above modeling would differ slightly if the animal died in a trap or was otherwise removed from the population in period 4. In this case, the capture history would be modeled as

$$\Pr(01010 \mid \text{release in period 2 and removal at period 4}) = \phi_2(1 - p_3)\phi_3 p_4$$

The χ_4 term is no longer appropriate as the final entry, because the animal is not exposed to survival and capture probabilities following period 4.

The complete likelihood for the CJS model can be written as the product of the probabilities associated with each capture history, conditional on the initial capture and release of each animal (Seber 1982; Lebreton et al. 1992). The result is a product-multinomial likelihood (a likelihood function resulting from multiplying more than one likelihood function, see chapter 1) that specifies a probability for each possible capture history given the number of animals exhibiting each capture history. Maximum likelihood estimation can then be used to obtain estimates of the model parameters, ϕ_j and p_j. Under the CJS model all parameters are time-specific. It is not possible to estimate the initial capture probability, p_1, and the final survival and capture probability of a study cannot be estimated separately, but only as a product $\phi_{k-1}p_k$.

The following assumptions typically are listed for the CJS model (e.g., Seber 1982; Pollock et al. 1990):

1. every marked animal present in the population at sampling period j has the same probability p_j of being recaptured or resighted;

2. every marked animal present in the population immediately following the sampling in period j has the same probability ϕ_j of survival until sampling period $j + 1$;

3. marks are neither lost nor overlooked, and are recorded correctly;

4. sampling periods are instantaneous (in reality they are very short periods) and recaptured animals are released immediately;

5. all emigration from the sampled area is permanent; and

6. the fate of each animal with respect to capture and survival is independent of the fate of any other animal.

Consequences of violating these assumptions have been well studied and are summarized in chapter 3 (also see Seber 1982, pp. 223–232; Pollock et al. 1990; Williams et al. 2002, pp. 434–436).

Substantial attention has been devoted in the literature to assumptions 1 and 2, which concern homogeneity of the rate parameters that underlie the capture-history data. Survival and capture probabilities frequently vary as a function of the attributes of a captured or observed animal. Much work in capture–recapture modeling over the last two decades has involved efforts to model variation in survival and capture parameters as functions of variables characterizing the state of individual animals at different times (e.g., Lebreton et al. 1992; Nichols et al. 1994).

Reduced-parameter Models

The CJS model yields closed-form estimators that are readily computed (chapter 3, this volume) and was virtually the only model used for estimating survival rates from capture–recapture data for the first 15–20 years following its development. However, with a separate parameter for capture and survival probability for each sampling period, the model was more general than needed in many cases, and this often resulted in estimates with large variances. As access to computers increased, statisticians began to consider reduced-parameter models that required iterative solutions to obtain parameter estimates (e.g., Cormack 1981; Sandland and Kirkwood 1981; Jolly 1982; Clobert et al. 1985; Crosbie and Manly 1985). As noted in chapter 3, the initial reduced-parameter models constrained either capture probability or survival probability or both parameters to be constant over time (i.e., $p_j = p$ and $\phi_j = \phi$).

In the event of unequal time intervals between sampling occasions, the constraint $\phi_j = \phi$ is not likely to be reasonable (i.e., survival probabilities will tend to be smaller for longer time intervals). In this case a more reasonable constraint would be for survival per unit time to be a constant, so that $\phi_j = \phi^{t_j}$, where t_j is the number of time units separating sampling occasions j and $j + 1$, and ϕ is the survival probability per unit time (Brownie et al. 1986).

The likelihoods for reduced-parameter models are created by substituting the single, time-constant parameter for the corresponding time-specific parameter in the CJS likelihood (e.g., p for p_j). Maximum likelihood estimation can then be conducted numerically (e.g., section 1.3; White and Burnham 1999), as closed-form estimators are not available. Likelihood ratio tests can be used to test between competing models (Lebreton et al. 1992), or information-theoretic approaches such as Akaike's Information Criterion (AIC) can be used to select the appropriate model from a set of candidate models (chapter 1; Akaike 1973; Anderson et al. 1994; Burnham and Anderson 1998).

Time-specific Covariates

In many situations, survival and capture probabilities can be modeled as functions of time-specific external variables. For example, environmental variables might influence either capture or survival probabilities,

or variables reflecting sampling effort might be related to capture probabilities. Such modeling involves incorporation of the hypothesized covariate model directly into the likelihood (North and Morgan 1979; Clobert
and Lebreton 1985; Clobert et al. 1987; Lebreton et al. 1992), and maximum likelihood estimation can then be used to estimate the parameters
of the functional relationship directly.

Lebreton et al. (1992) considered the use of a link function (McCullagh
and Nelder 1989), a function f that links the parameter of interest to a
(typically) linear function of (possibly) multiple covariates; for example,
$f(x_{gj})$, where x_{gj} denotes the value of covariate g at sampling occasion j,
and the β_g are parameters associated with this relationship. The link function is also often expressed using its inverse, f^{-1}. Link functions noted as
potentially useful by Lebreton et al. (1992) include the identity function
$f^{-1}(x) = x$, the logit function $f^{-1}(x) = \text{logit}(x) = \ln[x/(1 - x)]$, the logarithm
function $f^{-1}(x) = \ln(x)$, and the hazard function $f^{-1}(x) = \ln[-\ln(x)]$.

The logit link function is used frequently in capture–recapture modeling and associated software (e.g., White and Burnham 1999) and has the
advantages of providing a flexible form and bounded estimates for ϕ_j
and p_j in the interval $(0, 1)$. For example, modeling of survival using the
logit link essentially substitutes the following model for ϕ_j in the likelihood function:

$$\phi_j = \frac{\exp(\beta_0 + \beta_1 x_{1j} + \cdots + \beta_m x_{mj})}{1 + \exp(\beta_0 + \beta_1 x_{1j} + \cdots + \beta_m x_{mj})} \tag{5.1}$$

Maximum likelihood estimates of the β parameters can then be obtained
numerically via maximization of the likelihood function, using iterative
methods as described in chapter 1.

Multiple Groups

Capture–recapture data sometimes can be grouped into distinct cohorts.
For example, both males and females are captured in many studies, and it
is often reasonable to suspect that the sexes do not have equal survival
and recapture probabilities. One approach to parameter estimation would
be to use completely different parameters to model the capture histories
for the two sexes, estimating sex-specific survival and capture probabilities. For example, a three-period study would include the following pa-
rameters: $\phi_1^1, \phi_2^1, \phi_1^2, \phi_2^2, p_2^1, p_3^1, p_2^2, p_3^2$, where the superscript denotes the
sex (1 = male and 2 = female). Note that because the CJS and related
models are conditional on the initial capture (modeling involves only
subsequent events in the capture history), the above set of parameters does

not include the capture probabilities for the first sampling period (p_1^1, p_1^2). Also note that all of the above parameters cannot be separately estimated. In particular, under the CJS model, the final capture and survival probabilities cannot be estimated separately but only as a product ($\phi_{k-1} p_k$). Thus, although the example three-period model contains 8 parameters, only 6 can be estimated: $\phi_1^1, \phi_1^2, p_2^1, p_2^2, \phi_2^1 p_3^1, \phi_2^2 p_3^2$. Such a model is generally denoted as model (ϕ_{s*t}, p_{s*t}), where the model notation subscripts s and t denote sex and time, respectively. The asterisk indicates that if the model is written as a generalized linear model (McCullagh and Nelder 1989), it includes parameters for all interaction terms between the different levels of the associated factors s and t. For example, survival probabilities for the above three-period model can be written as

$$\phi_j = \frac{\exp(\beta_0 + \beta_1 I_s + \beta_2 I_{t(1)} + \beta_3 I_s I_{t(1)})}{1 + \exp(\beta_0 + \beta_1 I_s + \beta_2 I_{t(1)} + \beta_3 I_s I_{t(1)})} \qquad (5.2)$$

where β_0 is an overall intercept term, β_1 is a parameter associated with sex with $I_s = 1$ for males and 0 for females, β_2 is a parameter associated with the first sampling period 1 with $I_{t(1)} = 1$ when $j = 1$ and 0 otherwise, and β_3 is an interaction parameter associated with males in the first sampling period with $I_s = 1$ and $I_{t(1)} = 1$. Note that the survival model of equation (5.2) contains four parameters, as was also the case when we used a different survival parameter for each possible sex–time combination. These are simply two different ways of writing the same model.

In the case of general model (ϕ_{s*t}, p_{s*t}), none of the original parameters are the same for the two sexes. However, there are many situations in which either survival or capture probabilities or both should be similar for the two sexes. Such models can be constructed by imposing constraints on the general model. For example, the model (ϕ_{s*t}, p_t) includes different time-specific survival probabilities for males and females, yet assumes common capture probabilities for the sexes. In the above example with three sample periods, the total number of parameters (ignoring estimability questions) is reduced from 8 in the full model to 6 in model (ϕ_{s*t}, p_t).

A useful class of constraints for multiple groups involves the idea of parallelism (Lebreton et al. 1992), which links the temporal variation in a parameter for two or more groups. For example, Lebreton et al. (1992) use the notation (ϕ_{s+t}, p_{s*t}), to denote a model with time specificity and sex specificity of both survival and capture probability parameters, with the $s + t$ notation of the survival parameter indicating that survival varies over time, but does so in a parallel or additive manner for the two sexes. One way to write this model is to simply remove the interaction parameter from the full model of equation 5.2 to yield the parallel parameterization

$$\phi_j = \frac{\exp(\beta_0 + \beta_1 I_s + \beta_2 I_{t(1)})}{1 + \exp(\beta_0 + \beta_1 I_s + \beta_2 I_{t(1)})} \qquad (5.3)$$

The number of parameters is reduced by only one (from four to three) in this example, but in cases with moderate to large numbers of sample periods, the reduction can be substantial.

Finally, note that the modeling of multiple groups is very general and can include animals at multiple locations or perhaps exposed to multiple experimental treatments. Modeling of such data would then focus on the possible existence of spatial variation or treatment-induced variation, respectively, in the parameters of interest.

Effects of Capture History

As noted above, the CJS model assumes that all marked animals present in the population during sampling period j have equal probabilities of being caught (or resighted) that time, and of surviving to any subsequent sampling period. Robson (1969) and Pollock (1975) first considered more general models in which individuals alive in the sampled population at period j could exhibit different capture and survival probabilities at period j depending on their previous capture history.

A useful example of capture-history dependence involves trap response in capture probabilities (Cormack 1981; Sandland and Kirkwood 1981; Lebreton et al. 1992; Pradel 1993). Sandland and Kirkwood (1981) considered a simple model of trap dependence with different capture probabilities for an animal at period j, depending on whether or not the animal had been captured at period $j - 1$. That is,

p_j = the capture probability at sampling period j for an animal that was caught at $j - 1$; and

p'_j = the capture probability at sampling period j for an animal that was not caught at $j - 1$.

Sandland and Kirkwood (1981) obtained estimates under a reduced-parameter version of the above model in which the capture probabilities were constant over time, but survival probabilities were time dependent. Under this model, the subscript j is dropped so that $p_j = p$ and $p'_j = p'$.

To illustrate this parameterization, consider the probability of observing the following capture history 01010 under the above trap-dependence model of Sandland and Kirkwood (1981):

$$Pr(01010 \mid \text{release in period 2}) = \phi_2(1 - p)\phi_3 p'(1 - \phi_4 p)$$

The capture probability parameters associated with sampling periods 3 and 5 correspond to animals caught the previous time periods (2 and 4, respectively), whereas the capture probability for period 4 corresponds to an animal not caught the previous period.

Trap response can also occur in survival, such that survival between periods j and $j + 1$ depends on capture history before and including period j. For example, Brownie and Robson (1983) considered the sampling situation in which a mark is applied at initial capture, and subsequent encounters with marked animals are resightings. If trapping or handling adversely affects survival, then such an effect most likely occurs during the interval immediately following capture (the initial encounter). Brownie and Robson (1983) thus parameterized survival as

ϕ_j = the probability that a previously marked animal in the sampled population at time j survives until time $j + 1$ and remains in the sampled population; and

ϕ_j' = the probability that a previously unmarked animal in the sampled population at time j survives until time $j + 1$ and remains in the sampled population.

As an illustration of the parameterization under the Brownie-Robson (1983) model, consider the capture history and associated probability

$$Pr(01010 \mid \text{release in period 2}) = \phi_2'(1 - p_3)\phi_3 p_4(1 - \phi_4 p_5)$$

The survival probability following release in period 2 includes the prime notation because it corresponds to an animal that has not been previously marked, whereas the subsequent survival parameters correspond to resightings of marked animals. Both kinds of survival parameters can be written as linear-logistic functions of covariates as in equation 5.1.

Pradel et al. (1997a) reparameterized the Brownie-Robson (1983) model to correspond to the situation in which unmarked animals are viewed as being of two groups, residents that have a chance of surviving and being recaptured, and transients that are just moving through the study area and have no chance of being seen again on the area. The presence of transients (with apparent survival probability of 0) among unmarked animals causes the survival probability of new animals to be lower than that of marked animals (these are residents by definition). The parameterization of Pradel et al. (1997a) thus permits estimation of resident survival probability as well as the probability that an unmarked animal is a transient.

Individual Covariates

The CJS model and the models derived from it assume homogeneity of capture and survival probabilities among individuals at some level of grouping. For example, even in the case of dependence on previous capture history, homogeneity is assumed to apply to all animals with a particular history. Of course, strict homogeneity (exact equality) of survival and capture probabilities of different individual animals is unlikely ever to be true, regardless of the manner in which animals are grouped or categorized. Sometimes a substantial amount of variation among individuals in survival and capture probabilities may be explained by a single measurable covariate, for example: size or weight at some critical stage in the early life of an animal, mass at hatch or fledging (Perrins 1963, 1965), or parental size and experience (Hastings and Testa 1998).

Individual covariates of this type are static, in the sense that the single measurement characterizes the animal throughout the capture–recapture study (perhaps throughout life). Individual-based modeling is based on the view that the capture history of each individual animal is a multinomial sample with sample size of 1 (Smith et al. 1994). The full development of individual-based modeling is presented by Skalski et al. (1993), Smith et al. (1994), and Hoffman and Skalski (1995). An important point is that such modeling can be implemented in software such as SURPH (Smith et al. 1994) and MARK (White and Burnham 1999). As with multiple groups and time-specific covariates described above, link functions (e.g., logit or hazard) are used to model the individual survival or capture probabilities as functions of individual-level covariates, as well as group-specific and time-specific covariates.

We emphasize that these individual covariate models assume that the covariate characterizes the animal throughout the study. Time-varying individual covariates cannot be measured for animals that are not caught in a sampling period and thus cannot be used to model survival or capture probabilities. Such modeling requires models for the change in the covariate over time, and this is both reasonable and possible in some sampling situations (S. Bonner and C. Schwarz, personal communication). However, at present, use of time-varying individual covariates is most readily accomplished with existing software via discretization into groups and use of multistate models (Nichols et al. 1992b; chapter 8).

Model Selection and Related Issues

The large number of potential models, even for the single-age situation, emphasizes the need for a reasonable approach to model selection. Under one reasonable approach, the investigator begins with an a priori model set and must then decide which model(s) is best supported by the data, in the sense of adequately describing variation in the data in a parsimonious manner (without too many parameters). As noted in chapter 1, AIC (Akaike 1973) has been recommended for this purpose (e.g., Anderson et al. 1994; Burnham and Anderson 1992, 1998, 2002). Use of AIC presupposes that at least one model in the model set provides a reasonable fit to the data. Thus, a useful first step is to conduct a goodness-of-fit test of the most general model in the model set.

Goodness-of-fit tests have been developed specifically for the CJS model (Pollock et al. 1985, implemented in program JOLLY; Burnham et al. 1987, implemented in program RELEASE), and the Brownie-Robson trap response model (Brownie and Robson 1983, implemented in program JOLLY), for example. The CJS model tests of Pollock et al. (1985) and Burnham et al. (1987) have the same basic structure and differ only in their method of pooling of data. The CJS model test implemented in program RELEASE is labeled as the sum of two sets of contingency table tests, TEST 2, which focuses on subsequent capture histories of animals in different release cohorts, and TEST 3, which focuses on subsequent capture histories of animals within each release cohort, but with different prior histories (Burnham et al. 1987). The TEST 2 + TEST 3 sum is distributed as χ^2 under the null hypothesis of reasonable model fit. If it is judged to be significant, then the test statistic can be used to compute a variance inflation factor as

$$\hat{c} = \frac{\chi^2}{df}$$

where df is the degrees of freedom associated with χ^2 test statistic. Note that under the null hypothesis of reasonable model fit

$$E(\chi^2) = df$$

so that

$$E(\hat{c}) = 1$$

For other models without well-developed goodness-of-fit tests, a Pearson goodness-of-fit test can often be computed. However, in many cases these tests do not perform well when a large proportion of contingency table cells must be pooled because of sparse data and low expected cell values. An alternative means of assessing fit is based on a parametric bootstrap approach that can be implemented using program MARK (White et al. 2001). This method uses the estimates for the model of interest, which is frequently the most general model in the model set. The estimates from this model for the real data are assumed to be good approximations for the true parameter values, and many new data sets are generated by simulation using the model with the estimated parameter values. The assumed model is fitted to each of the generated sets of data, and the deviance is computed. The observed deviance from the actual capture history data set can then be compared with the generated distribution of deviances to assess fit.

This then yields two approaches for estimating $\hat{c}$ (White et al. 2001). Under the first approach, it is estimated as the observed deviance divided by the mean of the bootstrap deviances, as this latter mean should estimate the deviance in the case where the model fits the data adequately. The second approach uses the observed $\hat{c}$ (computed as the deviance divided by the model df) divided by the mean of the deviance/df from the bootstrap iterations. These two approaches give different results because the df for a set of data varies from one generated data set to the next with parametric bootstrapping. Neither of the approaches works well in all situations, and goodness of fit is an important issue requiring additional work in open capture–recapture modeling.

Most of the theory underlying use of such a variance inflation factor is based on the assumption that overdispersion is the source of the lack of fit, but the $\hat{c}$ factor is used more widely, as the true reasons underlying a lack of fit are seldom known. In the case of overdispersion, a model-based variance estimate $\text{v\^{a}r}(\hat{\theta})$ will typically be too small and should be modified to $\hat{c} \, \text{v\^{a}r}(\hat{\theta})$.

If even the most general model in the set does not appear to fit the data well, then a quasilikelihood approach is recommended for both likelihood ratio testing and model selection (Burnham et al. 1987; McCullagh and Nelder 1989; Lebreton et al. 1992; Burnham and Anderson 2002, chapter 1). In the case of overdispersion, likelihood ratio tests can be modified using $\hat{c}$ to yield F statistics for testing between nested models (Lebreton et al. 1992). In the case of lack of fit, or when $\hat{c} > 1$, the AIC model selection statistic can be modified to

$$\text{QAIC} = -\left[2\ln\left(\frac{L(\hat{\theta})}{\hat{c}} \right) \right] + 2K$$

where $L(\hat{\theta})$ is the likelihood function evaluated at the MLEs of the parameters, and K is the number of parameters in the model, plus one parameter corresponding to $\hat{c}$ (Burnham and Anderson 2002). In addition to this quasilikelihood adjustment, most general capture–recapture models have large numbers of parameters to be estimated from relatively small samples, leading to an AIC statistic adjusted for both quasi-likelihood and small sample size:

$$\mathrm{QAIC_c} = -\left[2\ln\left(\frac{L(\hat{\theta})}{\hat{c}}\right)\right] + 2K + \frac{2K(K+1)}{n-K-1}$$

where n denotes the effective sample size. The sample size is not so easily defined in capture–recapture modeling, and Burnham and Anderson (2002) suggest use of the number of distinct animals captured once or more.

Model selection proceeds using ranked AIC or $\mathrm{QAIC_c}$ values, with the model exhibiting the lowest value selected as the most appropriate for the data. Model ranks based on this selection process are typically expressed as $\Delta\mathrm{AIC}_i$, or simply as Δ_i, computed as the difference between the AIC statistic for model i and the model with minimum AIC. These Δ_i can then be used to compute model-specific AIC weights:

$$w_i = \frac{\exp(-\Delta_i/2)}{\sum \exp(-\Delta_r/2)}$$

where the summation is over all of the models in the model set. These weights reflect the proportional support or weight of evidence for a specific model (Burnham and Anderson 2002). If several models exhibit similar AIC values, then weighted estimates of important parameters can be obtained as a weighted mean of estimates from multiple models, with each model's estimate being weighted by its AIC weight (Buckland et al. 1997; Burnham and Anderson 2002).

Example

Methods for single-age capture–recapture analysis are illustrated here with a live-trapping study of meadow voles, *Microtus pennsylvanicus*, at the Patuxent Wildlife Research Center, Laurel, MD. Details of trapping and field methods are provided by Nichols et al. (1984b). Although trapping was conducted under the robust design (Pollock 1982), the five days of sampling in each month were collapsed into a single assessment of whether each animal was caught at least once, or not, for each of the six monthly sampling occasions. Resulting capture history data are presented in table 5.1. Note that a small number of animals was lost on capture because they died in the trap.

TABLE 5.1

Adult capture history data for a six-period study of meadow voles, *Microtus pennsylvanicus*, at Patuxent Wildlife Research Center, Laurel, Maryland, 1981

Capture history	Number of females	Number of males
100000	7	8
100000	−1	−3
110000	10	21
110000	−4	−2
111000	7	5
111100	2	3
111100	0	−1
111110	1	1
111110	−1	0
111111	14	10
101111	1	0
110111	1	1
101011	1	0
101110	1	0
101100	0	1
101000	1	0
10000	2	10
10000	0	−1
11000	2	3
11100	3	3
11110	1	0
11111	3	3
11101	0	3
11011	0	1
10111	1	2
10100	1	1
10001	0	1
1000	7	7
1000	0	−1
1100	2	3
1110	1	0
1111	4	3
1111	0	−1
100	6	13
110	0	2
110	0	−1
111	2	4
101	0	1
10	5	4
11	18	13
1	27	34

Note. Numbers preceded by "−" indicate animals that were not released following their last capture.

TABLE 5.2

Parameter estimates under the general two-sex CJS model (ϕ_{s*t}, p_{s*t}) for adult meadow voles studied at Patuxent Wildlife Research Center, Laurel, Maryland, 1981

| Capture period (j) | Sampling dates | Capture probability | | Survival probability | |
| | | Female | Male | Female | Male |
		$\hat{p}_j$ ($\hat{SE}[\hat{p}_j]$)	$\hat{p}_j$ ($\hat{SE}[\hat{p}_j]$)	$\hat{\phi}_j$ ($\hat{SE}[\hat{\phi}_j]$)	$\hat{\phi}_j$ ($\hat{SE}[\hat{\phi}_j]$)
1	27 Jun–1 Jul	—[a]	—[a]	0.89(0.052)	0.86(0.052)
2	1 Aug–5 Aug	0.88(0.055)	0.96(0.039)	0.78(0.066)	0.58(0.066)
3	29 Aug–2 Sep	0.90(0.057)	0.82(0.071)	0.68(0.066)	0.71(0.072)
4	3 Oct–7 Oct	0.96(0.037)	0.91(0.059)	0.69(0.069)	0.59(0.069)
5	31 Oct–4 Nov	1.00[b]	0.83(0.069)	—[a]	—[a]
6	4 Dec–8 Dec	—[a]	—[a]	—[a]	—[a]

[a] Parameter not estimable under CJS model.
[b] Standard error not estimated.

The most general model considered (ϕ_{s*t}, p_{s*t}) for fitting to the data represents a combination of separate CJS models for the two sexes. The fit of this model to the data was judged to be adequate ($\chi^2_{18} = 24.6$, $P = 0.14$) based on the overall goodness-of-fit test of program RELEASE (Burnham et al. 1987). The high capture probability estimates under this model reflect the five days of trapping, and the monthly survival estimates are typical of meadow voles (table 5.2).

A number of reduced-parameter models were also fitted to these data. For a full analysis see Williams et al. (2002, pp. 436–439). Several of these models were judged to be more appropriate than (ϕ_{s*t}, p_{s*t}). The general model (ϕ_{s*t}, p_{s*t}) showed $\Delta AIC_c = 5.25$ relative to the low-AIC_c model, (ϕ_{s+t}, p). Under the low-AIC_c model (ϕ_{s+t}, p), survival probability varied by sex and time, but the temporal variation was parallel (on a logit scale) for the two sexes (figure 5.2), with survival for females slightly higher than that for males. The capture probability was best modeled using a single value ($\hat{p} = 0.90$, $\hat{SE}(\hat{p}) = 0.020$) that was constant over time and the same for both sexes.

5.3 Conditional Multiple-age Models

In addition to the models described above, one way to relax the CJS assumption of homogeneous capture and survival probabilities is to permit age-specific variation. There are three classes of age-specific models that differ in the data structures for which they were developed.

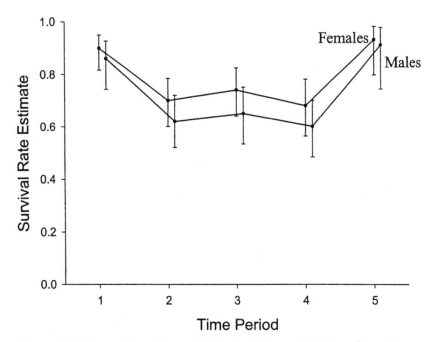

Figure 5.2. Estimated monthly survival probabilities and 95% confidence intervals from model (ϕ_{s+t}, p) for male and female meadow voles at Patuxent Wildlife Research Center.

Pollock's Multiple-age Model

This model was developed by Pollock (1981b); also see Stokes (1984). It assumes the existence of $I + 1$ age classes ($0, 1, \ldots, I$) that can be distinguished for newly caught (unmarked) animals, with age class I including all animals of at least age I. The model requires a design restriction that the timing of sampling and age class transition are synchronized, such that an individual of age v in sample period j will be at age $v + 1$ in sample period $j + 1$. Here, we discuss the simplest case in which $I = 1$, with young ($v = 0$) and adults ($v = 1$) as the distinguishable age classes. Estimation under the Pollock (1981b) model is based on the numbers of animals in each age class exhibiting each of the observable capture histories (denoted by $x_\omega^{(v)}$ for capture history ω and age v, where v corresponds to the age at initial capture). For example, $x_{011}^{(0)}$ is the number of animals released as young in period 2 and recaptured (as adults) in period 3.

Parameters are defined in a manner similar to the single-age case, with the additional notation of a superscript denoting age:

$p_j^{(v)}$ = the probability that a marked animal of age v in the study population at sampling period j is captured or observed during period j;

$\phi_j^{(v)}$ = the probability that a marked animal of age v in the study population at sampling period j survives until period $j+1$ (to age $v+1$) and remains in the population (does not permanently emigrate); and

$\chi_j^{(v)}$ = the probability that an animal of age v in the study population at sampling period j is not caught or observed again at any sampling period after period j. As in the case of single-age models, the $\chi_j^{(v)}$ parameters are written as functions of $p_j^{(v)}$ and $\phi_j^{(v)}$ parameters. In the two-age case, for example, $\chi_j^{(1)}$ is defined just as in the CJS model using all adult parameters, and $\chi_j^{(0)}$ can be written as

$$\chi_j^{(0)} = (1 - \phi_j^{(0)}) + \phi_j^{(0)}(1 - p_{j+1}^{(1)})\chi_{j+1}^{(1)}$$

The modeling of capture-history data proceeds in the same manner as with the single-age CJS model. Consider capture history 01010 for animals marked in period 2. Modeling of this history is again conditional on the initial capture at period 2 and is dependent on the age of initial marking. In the 2-age situation,

$$\text{Pr}(01010 \mid \text{release in period 2 as young}) = \phi_2^{(0)}(1 - p_3^{(1)})\phi_3^{(1)}p_4^{(1)}\chi_4^{(1)}$$
$$\text{Pr}(01010 \mid \text{release in period 2 as adult}) = \phi_2^{(1)}(1 - p_3^{(1)})\phi_3^{(1)}p_4^{(1)}\chi_4^{(1)}$$

These probabilities differ only in the superscript of the initial survival probability. Animals initially released as young survive the interval following initial release (periods 2–3) with a survival parameter associated with young animals. But subsequent capture and survival probabilities correspond to adult animals, as the young animal in period 2 makes the transition to the adult class in the interval between periods 2 and 3. As with single-age models, if the animal is removed (not released) following the last capture, then the final $\chi_j^{(v)}$ term is simply removed from the capture history model. Age-specific modeling of survival can also be accomplished by writing the logit of survival as a linear function of age in the same way as was done for sex, i.e., by substituting I_a (1 for adults and 0 for young) for I_s, and $I_{a(tj)}$ for $I_{s(tj)}$ in equation 5.2. As was the case for the additive modeling of sex and time effects, deletion of the interaction terms results in a model in which survival differs for the different ages but is parallel (on a logit scale) over time.

The complete likelihood for the Pollock (1981b) multiple-age model can be written as a product-multinomial likelihood that specifies a probability for each possible capture history together with the actual data, the number of animals exhibiting each capture history. Maximum likelihood estimation can then be used to obtain estimates of the model parameters, $\phi_j^{(v)}$ and $p_j^{(v)}$.

Model assumptions are very similar to those listed for the single-age CJS model. Assumptions 1 and 2 of the CJS model are modified for the age-specific model to restrict homogeneity of capture $(p_j^{(v)})$ and survival $(\phi_j^{(v)})$ probabilities to members of the same age class (v) at each sampling period (e.g., survival probability must be the same for all animals of age v but not for animals of different age classes). The age-specific models of Pollock (1981b), Stokes (1984), and Brownie et al. (1986) also assume that age is correctly assigned to each new animal that is encountered and marked.

Reduced-parameter models were presented for multiple age classes by Brownie et al. (1986) and Clobert et al. (1987), and are also described in Pollock et al. (1990) and Lebreton et al. (1992). Reduced-parameter models of interest include those constraining parameters to be constant over time, as well as models constraining parameters to be constant across age classes. Time-specific and individual covariate models can be developed for multiple ages following the same general principles outlined for single-age models. Models with multiple groups and capture-history dependence are also developed for multiple ages in the same manner as for a single age. Estimation under these various models can be carried out by program MARK (White and Burnham 1999). Finally, the general principles of model selection described for single-age models are also applicable to multiple ages. Goodness-of-fit tests for the Pollock (1981b) and reduced-parameter models were developed by Brownie et al. (1986); also see Pollock et al. (1990).

Age 0 Cohort Models

It is also possible to focus on cohorts of animals released as newborns (age 0) in situations where the age of organisms can be distinguished only in terms of young (age 0) and older (age >0) individuals. The only way to know the specific age of an adult animal is for the animal to have been released in a previous period at age 0. New, unmarked adults are thus not used in the modeling. The development of these models assumes that the interval between sampling occasions coincides with the time period required for the animals to mature from one age class into the next (e.g., annual sampling with year age classes). The notation for capture-history data is the same as that for the Pollock (1981b) model. With the cohort model, however, all animals used for modeling are marked as young ($v = 0$), so all capture-history statistics are superscripted with (0). Cohort models also have been used for unaged adults (Loery et al. 1987). In such cases, the superscript for both data and parameters corresponds to the number of time periods since initial capture rather than to age. This notation highlights the operational definition of cohort in the context of these models as a group of animals initially captured at the same occasion and, in some cases, of the same age.

Models for cohort data were considered by Buckland (1980, 1982), Loery et al. (1987), and Pollock et al. (1990). Notation is similar to that of the age-specific models of Pollock (1981b), with probabilities of capture ($p_j^{(v)}$) and survival ($\phi_j^{(v)}$). Modeling is also similar to that for the Pollock (1981b) models, except that age is defined not only for classes recognizable at capture but for animals of all ages, given that they were initially caught at age $v = 0$. For example, the probability associated with capture history 01010 for individuals first captured as young in year 2 is

$$\Pr(01010 \mid \text{release at period 2 as young}) = \phi_2^{(0)}(1 - p_3^{(1)})\phi_3^{(1)}p_4^{(2)}\chi_4^{(2)}$$

Every increase in sample period (subscript j) is accompanied by an increase in age (superscript v). This general cohort model can be viewed as a series of separate CJS models, one model for each cohort of age-0 releases. Age-0 cohort models and various derivative models (e.g., reduced-parameter, covariates, multiple groups) can be implemented in MARK (White and Burnham 1999). This class of models can be used to address questions about senescent decline in survival probability (see Pugesek et al. 1995; Nichols et al. 1997). For such questions it is frequently useful to write the logit of age-specific survival rate as a linear function of age.

Age-specific Breeding Models

Not all ages are exposed to sampling efforts under some capture–recapture sampling designs. Young of many colonial breeding bird species depart the breeding ground of origin following fledging and do not return to the breeding colony of origin until they are ready to breed. Thus, prebreeders of age >0 can be viewed as temporary emigrants with probability 0 of being captured or observed prior to their first breeding attempt. Temporary emigration of this sort can be dealt with using either the robust design (see section 5.6) or standard open-model capture history data modeled with a structure that accommodates the absence of prebreeders.

Using this latter approach, Rothery (1983) and Nichols et al. (1990) considered estimation in the situation where all birds begin breeding at the same age. Clobert et al. (1990, 1994) considered the more general situation where not all animals begin breeding at the same age. Clobert et al. (1994) describe a two-step approach in which age-specific breeding probabilities are estimated as functions of capture probabilities estimated from a cohort model. Spendelow et al. (2002) and Williams et al. (2002, pp. 447–454) used this idea to develop a model that estimates the probability that an animal of age v that has not become a breeder before time j does so at time j. Although the general model has been used primarily for birds, it may be useful for a variety of other groups, including sea turtles, anadromous fish, some amphibians, and some marine mammals.

5.4 Reverse-time Models

The models discussed thus far are conditioned on numbers of releases at a given sampling period, and they describe the remainder of the capture history. Pollock et al. (1974) noted that if the capture-history data are considered in reverse time order, conditioning on animals caught in later time periods and observing their captures in earlier occasions, then inference can be made about the recruitment process. More recent uses of reverse-time capture–recapture modeling include Nichols et al. (1986, 1998), Pradel (1996), Pradel et al. (1997b) and Pradel and Lebreton (1999).

The data for reverse-time modeling are the same as previously described for forward-time modeling and consist of the numbers of animals exhibiting each observable capture history. The modeling is similar to that described above for the CJS model, the time direction being the only real difference. Two primary parameters are defined to be

γ_j = the probability that an animal present just before time j was present in the sampled population just after sampling at time $j-1$, and

p'_j = the probability that an animal present just after sampling at time j was captured at j.

The parameters γ_j were termed "seniority" parameters by Pradel (1996). They can be viewed as survival probabilities that extend backward in time. Note that both the seniority and capture probabilities are defined carefully relative to the time of sampling (just before or after sampling). The reason for this attention to timing concerns losses on capture. For the discussion here, we will typically assume no losses on capture for ease of presentation. In this case, the capture probability parameters are identical for forward and reverse time modeling, $p'_j = p_j$.

In addition to the above parameters, let ξ_j be the probability of not being seen previous to time j for an animal present immediately before j.

The parameter ξ_j, which is analogous to χ_j in standard-time modeling, can be written recursively

$$\xi_j = (1 - \gamma_j) + \gamma_j(1 - p'_{j-1})\xi_{j-1}$$

for $j = 2, 3, \ldots, k$, with $\xi_1 = 1$. To have not been seen before time j, an animal either must not be a survivor from time $j - 1$ (this possibility occurs with probability $1 - \gamma_j$), or it must be a survivor (with probability γ_j) that was not caught in $j - 1$ (with probability $1 - p'_{j-1}$) and not seen before $j - 1$ (with probability ξ_{j-1}).

Consider the reverse-time modeling of capture history data, using the same history that was used to illustrate standard-time modeling. Again consider history 01010, indicating capture in periods 2 and 4 of a five-period study. For reverse-time modeling we condition on the final capture and model prior events in the capture history:

$$\Pr(01010 \mid \text{last capture at period 4}) = \gamma_4(1 - p'_3)\gamma_3 p'_2 \xi_2$$

Beginning with the final capture in period 4 and working backward, the animal exhibiting this history was an old animal at 4, in the sense that it was a survivor from period 3. The associated probability is γ_4. The animal was not captured at time 3 (associated probability is $1 - p'_3$). It was a survivor from 2 (γ_3), and it was again caught at 2 (p'_2). However, it was not seen before period 2 (ξ_2). Note that unlike the case with standard-time modeling, the reverse-time capture history modeling does not differ depending on whether or not the animal was released following the final capture in the history. The reverse-time modeling only involves events occurring prior to this time.

Conditional multinomial models can be developed by conditioning on the number of animals caught for the last time at each period, j, and then using the numbers of these animals exhibiting each capture history, in conjunction with the probabilities associated with each history. Estimation of model parameters is accomplished using the method of maximum likelihood. In fact, estimates of γ_j and p'_j can be obtained by simply reversing the time order of capture-history data and obtaining estimates using software developed for standard-time analyses (Pradel 1996). Program MARK (White and Burnham 1999) contains a routine to provide estimates under reverse-time modeling.

The assumptions underlying reverse-time modeling are similar to those underlying standard-time models. The homogeneity assumptions (1,2) now apply to the seniority and capture probabilities rather than to survival and capture probabilities. For reverse-time modeling, every marked and unmarked animal present in the population at sampling period j

must have the same probability p'_j of being captured. This assumption can be restrictive and is important to recall when estimating γ_j.

The reverse-time parameters, γ_j, reflect the proportional change in population growth rate, λ_j, that would result from a specified proportional change in survival probability. The complement, $1 - \gamma_j$, reflects the proportional change in λ_j that would result from a specified proportional change in recruitment of new animals. These parameters are thus analogous to the elasticities computed for projection matrix parameters (Caswell 2001). A discussion of the use and interpretation of reverse-time capture–recapture estimates is presented by Nichols et al. (2000). Combination of standard-time and reverse-time modeling in a single likelihood permits direct estimation of λ_j and is considered in section 5.5, as well as by Pradel (1996) and Nichols and Hines (2002).

The same kinds of modeling can be used with reverse-time capture–recapture as with standard time. Reduced parameter models and models with time-specific and individual covariates can be developed. Multiple age models can also be developed but require multistate models (Arnason 1973; Hestbeck et al. 1991; Brownie et al. 1993; chapter 8) and a robust design approach (see Nichols et al. 2000).

5.5 Unconditional Models

Here, we consider the estimation of population size and recruitment using capture–recapture data for open (to gains and losses between sampling occasions) populations. We focus on single-age models, where we consider all animals as adults.

For standard capture–recapture sampling, the data collected are identical to those used for the conditional models described above, but there is increased emphasis on the number of unmarked animals in unconditional modeling. In standard capture–recapture studies, unmarked animals that are captured are given tags permitting individual identification, and the number of these animals is recorded. In capture–resighting studies, an effort must be made to count the number of unmarked animals encountered during the resighting sampling efforts. These counts of unmarked animals, which are not needed for survival rate estimation with conditional models, play a key role in estimation of population size and recruitment.

The Jolly-Seber (JS) Model

The JS model includes the CJS structure for modeling capture histories conditional on initial capture and release. The JS model also includes an additional component for modeling the number of unmarked animals caught in sampling period j (denote as u_j) as a function of the total number of unmarked animals, U_j, in the population, i.e., $E(u_j | U_j) = p_j U_j$.

The following are unknown random variables, the values of which are to be estimated:

N_j = the total number of animals in the population exposed to sampling efforts in sampling period j; and

B_j = the number of new animals joining the population between samples j and $j + 1$, and present at $j + 1$.

Jolly (1965) and Seber (1965) presented closed-form estimators of the above quantities, as discussed in chapter 3.

These estimators can be derived by obtaining maximum likelihood estimates of survival and capture probabilities using the conditional portion of the likelihood, which is the same as the entire CJS likelihood (Brownie et al. 1986; Williams et al. 2002), and then applying the resulting time-specific estimates of capture probability to the total number of animals caught in each sampling period. Thus, $n_j = u_j + m_j$, where m_j denotes the number of marked animals caught at j, and

$$\hat{N}_j = \frac{n_j}{\hat{p}_j} \tag{5.4}$$

Substitution of $\hat{p}_j = m_j / \hat{M}_j$ (where $\hat{M}_j$ is the estimated number of marked animals in the population just before sampling period j) into equation 5.4 yields the JS abundance estimator

$$\hat{N}_j = \frac{n_j \hat{M}_j}{m_j}$$

The assumptions listed for the CJS model also are required for the JS model. However, assumption 1, that every marked animal in the population at sampling period j has the same probability of being recaptured or resighted, must be modified for JS application to state that the capture probabilities p_j also apply to unmarked animals. This requirement follows from the estimation of capture probability from data on marked animals and yet the need to apply this capture probability to both marked and unmarked animals to estimate abundance.

Additional unconditional models can be considered following the general JS modeling approach. Partially open models, open only to losses from, or only to gains to, the population, were developed by Darroch (1959) and Jolly (1965) and are discussed in chapter 3 of this volume. Abundance can also be estimated under other reduced-parameter models,

models with time-specific covariates, multiple-age models, multiple-group models, multistate models, and some models with capture-history dependence. See the review in Williams et al. (2002). Although specific estimators have been developed for some reduced-parameter and multiple-age models (e.g., Pollock 1981b; Jolly 1982; Brownie et al. 1986), one general approach to estimation involves use of equation 5.4 in conjunction with the appropriate n_j and $\hat{p}_j$.

If the capture probability is estimated at the individual level based on covariates, then the animals captured on occasion j represent a heterogeneous mixture of capture probabilities and equation 5.4 cannot be applied directly. A reasonable approach to estimation in this situation involves an estimator of the type proposed by Horvitz and Thompson (1952) and used in capture–recapture for closed models by Huggins (1989) and Alho (1990). Recently the approach was proposed by McDonald and Amstrup (2001) for use with open models.

Let I_{ji} be an indicator variable that assumes a value of 1 if animal i is captured in sampling period j, and 0 if the animal is not caught during j. Let $\hat{p}_{ji}$ be the estimated capture probability for animal i in period j, based on covariates associated with animal i and on an assumed relationship between capture probability and the relevant covariates. Abundance at period j then can be estimated as

$$\hat{N}_j = \sum_{i=1}^{n_j} \frac{I_{ji}}{\hat{p}_{ji}} = \sum_{i=1}^{n_j} \frac{1}{\hat{p}_{ji}} \qquad (5.5)$$

Note that if all animals have the same value of the relevant covariate (i.e., if there is no heterogeneity), then equation 5.5 reduces to equation 5.4. McDonald and Amstrup (2001) investigated the properties of this estimator using simulation, and concluded that it exhibited little bias.

Finally, we note that model selection for unconditional models follows the basic approach outlined under conditional models (also see chapter 1). As most of the information available for modeling capture probability comes from recaptures of marked animals, model selection and model fit are generally based on this portion of the likelihood.

Superpopulation Approach

Crosbie and Manly (1985) and Schwarz and Arnason (1996) reparameterized the Jolly-Seber model by directing attention to a new parameter, N, denoting the size of a superpopulation, so that N can be thought of as either the total number of animals available for capture at any time during the study or, alternatively, as the total number of animals ever in the sampled area between the first and last sampling occasions of the study. Space precludes development of the superpopulation approach here, but it can be noted that the POPAN software permits modeling and estimation (Arnason and Schwarz 1999). This approach has been used to estimate the number of salmon returning to a river to spawn (Schwarz et al. 1993b; Schwarz and Arnason 1996), and has been recommended as a means of estimating the number of birds passing through a migration stopover site (Nichols and Kaiser 1999).

Temporal Symmetry Approach

The temporal symmetry of capture–recapture data noted above was used by Pradel (1996) to develop an approach that included both standard-time and reverse-time parameters in the same likelihood. In particular, Pradel (1996) noted that this approach permits direct modeling of the population growth rate, $\lambda_j = N_{j+1}/N_j$, where N_j is again the abundance at sampling period j. The appearance of population growth rate as a model parameter can be understood by considering two alternative ways of writing the expected number of animals alive in two successive sampling occasions. This expectation can be written as $N_j \phi_j$ (the animals at j that survive until $j + 1$), based on forward-time modeling, and as $N_{j+1}\gamma_{j+1}$ (the animals at $j + 1$ that were members of the population at j), based on reverse-time modeling. Solving this equality yields an expression for the population growth rate of

$$\lambda_j = \frac{N_{j+1}}{N_j} \approx \frac{\phi_j}{\gamma_{j+1}} \qquad (5.6)$$

Thus, the expected number of animals exhibiting capture history 01010 under Pradel's (1996) temporal symmetry model can be written as

$$E(x_{01010}|N_1) = N_1 \left(\frac{\phi_1}{\gamma_2}\right)\xi_2 p_2 \phi_2 (1 - p_3)\phi_3 p_4 \chi_4 \qquad (5.7)$$

where x_{01010} is the number of animals with capture history 01010. The term $N_1(\phi_1/\gamma_2) = N_1\lambda_1$ gives the expected number of animals in the population just before sampling period 2, and ξ_2 is the probability that an animal in this group was not caught prior to sampling period 2. The animals exhibiting this history were caught at period 2, and the associated probability is p_2. The subsequent (for sample periods later than 2) modeling is similar to that presented above for the CJS model.

Expectations for the numbers of animals exhibiting each possible capture history can be written as in equation 5.7, conditioned on N_1. However, such expectations do not lead directly to a probability distribution, because they all contain N_1, an unknown random variable. Let x_h be the number of animals exhibiting capture history h, and M denote the total number of animals caught in the entire study, so that $M = \Sigma x_h$. The expected number of animals caught during a study can be written as the sum of the expected number of animals caught for the first time at each sampling occasion, so that

$$E(M) = \sum_{j=1}^{k} \xi_j N_j p_j = N_1 \sum_{j=1}^{k} \xi_j p_j \left[\prod_{i=1}^{j-1} \left(\frac{\phi_i}{\gamma_{i+1}} \right) \right] \tag{5.8}$$

Finally, the conditional probability, conditioned on the total of M animals caught, associated with a particular capture history, can be obtained by dividing the expected number of animals with that history, as in equation 5.7, by the expected number of total individual animals caught during the study, as in equation 5.8:

$$P(h) = \frac{E(x_h)}{E(M)} \tag{5.9}$$

The initial population sizes in the numerator and denominator of equation 5.9 cancel, leaving the conditional probabilities of interest expressed in terms of estimable model parameters. Then the likelihood L for the set of animals observed in a study can be written as the product of the conditional probabilities associated with all the individual capture histories (Pradel 1996):

$$L = \prod_{h} P(h)^{x_h} \tag{5.10}$$

Pradel (1996) suggested three different parameterizations for the above likelihood, each of which might be useful in addressing specific questions and all of which retain capture (p_j) and survival (ϕ_j) probabilities. The model described above is written in terms of p_j, ϕ_j, and γ_j. A second parameterization is based on p_j, ϕ_j, and λ_j, and is obtained by substituting the following expression for the γ_j of the original parameterization:

$$\gamma_j = \frac{\phi_{j-1}}{\lambda_{j-1}}$$

A third parameterization is based on a measure f_j of recruitment rate, which denotes the number of recruits to the population at time $j + 1$ per animal present in the population at j. Because $\lambda_j = \phi_j + f_j$, the f_j parameterization can be obtained via the following substitution of

$$\gamma_j = \frac{\phi_{j-1}}{\phi_{j-1} + f_{j-1}}$$

for γ_j of the original parameterization.

Potential uses of these different parameterizations are discussed by Nichols and Hines (2002) and Williams et al. (2002, pp. 514–515). Modeling and estimation under all three parameterizations are incorporated into program MARK (White and Burnham 1999). The basic assumptions are the same as for the JS and superpopulation approaches and are discussed with respect to this set of models by Hines and Nichols (2002).

Example

The meadow vole data of table 5.1 were used to estimate abundance, N_j, and recruitment, B_j, using the closed-form, bias-adjusted estimators presented for the JS model by Seber (1982) and in chapter 3. Estimates were precise, as expected based on the high capture probabilities, and indicated a total (adult males and females) fall population of about 110–135 animals (table 5.3).

These data were also used with Pradel's (1996) general temporal symmetry model with both γ (ϕ_t, p_t, γ_t) and λ (ϕ_t, p_t, λ_t) parameterizations to estimate both the proportion of animals at each period that were survivors from the previous period, γ_j, and the monthly population growth rate, λ_j (table 5.4). Estimates were obtained from a program written by James Hines, rather than from program MARK, because the latter program does not implement Pradel's (1996) suggested means of dealing with losses on capture, and such losses occurred in these data (table 5.1). However, losses on capture were small, so the estimates obtained using program MARK are very similar to those in table 5.4.

The estimates of the γ's indicate that about 60–80% of the adult population in any month represented survivors from the previous month, with the remaining 20–40% representing new recruits (table 5.4). The estimates of the λ's range from approximate 20% declines to 20% increases in abundance, but these estimates have poor precision under this general model. An improvement in the precision of the estimates of λ's can be obtained by imposing constraints on them, such as making them all equal. Finally, note that the parameter estimates in table 5.4 do not exactly equal

TABLE 5.3

Abundance and recruitment estimates under the JS model for adult meadow voles studied at Patuxent Wildlife Research Center, Laurel, Maryland, 1981

		Abundance		Recruitment	
		Female	Male	Female	Male
Capture period (j)	Sampling dates	$\hat{N}_j$ ($\hat{SE}[\hat{N}_j]$)	$\hat{N}_j$ ($\hat{SE}[\hat{N}_j]$)	$\hat{B}_j$ ($\hat{SE}[\hat{B}_j]$)	$\hat{B}_j$ ($\hat{SE}[\hat{B}_j]$)
1	27 Jun–1 Jul	—[a]	—[a]	—[a]	—[a]
2	1 Aug–5 Aug	60(2.35)	74(2.14)	14(2.07)	17(2.56)
3	29 Aug–2 Sep	58(2.51)	59(3.53)	7(1.35)	21(2.71)
4	3 Oct–7 Oct	47(1.09)	62(3.01)	23(0.48)	19(2.82)
5	31 Oct–4 Nov	55[b]	55(2.96)	—[a]	—[a]
6	4 Dec–8 Dec	—[a]	—[a]	—[a]	—[a]

[a] Parameter not estimable under Jolly-Seber model.
[b] Standard error not estimated.

the estimates of the same parameters derived from the JS estimates in table 5.3. There are two reasons for this. First, there is the use of bias-adjusted estimates in table 5.3. Second, there is the fact that the ratios of population size estimates reflect changes in abundance resulting from both trap loss and natural processes, whereas the estimates in table 5.4 are restricted to natural processes. See the discussion in Nichols and Hines (2002).

TABLE 5.4

Estimated seniority parameters (γ_j) and population growth rates (λ_j) under the temporal symmetry models (ϕ_t, p_t, γ_t) and (ϕ_t, p_t, λ_t) of Pradel (1996) for capture–recapture data on adult male and female meadow voles at Patuxent Wildlife Research Center

	Females		Males	
Capture period	$\hat{\gamma}_j$ ($\hat{SE}(\hat{\gamma}_j)$)	$\hat{\lambda}_j$ ($\hat{SE}(\hat{\lambda}_j)$)	$\hat{\gamma}_j$ ($\hat{SE}(\hat{\gamma}_j)$)	$\hat{\lambda}_j$ ($\hat{SE}(\hat{\lambda}_j)$)
1	—[a]	—[a]	—[a]	—[a]
2	—[a]	1.04 (0.118)	—[a]	0.83 (0.105)
3	0.75 (0.064)	0.80 (0.095)	0.71 (0.068)	1.07 (0.139)
4	0.85 (0.061)	1.18 (0.333)	0.67 (0.070)	0.90 (0.124)
5	0.59 (0.089)	—[a]	0.65 (0.075)	—[a]
6	0.63 (0.079)	—[a]	0.60 (0.064)	—[a]

[a] Quantity not estimable under the model.

5.6 The Robust Design

The JS model has been widely used in animal population ecology, but it has long been known that estimates of survival probability are more robust to deviations from model assumptions than estimates of abundance (e.g., see Carothers 1973b; Gilbert 1973). As a means of dealing with this difference in robustness, Pollock (1981a, 1982) suggested sampling at two temporal scales, with periods of short-term sampling over which the population is assumed to be closed, and longer-term sampling over which gains and losses are expected to occur (Lefebvre et al. 1982). He recommended that the closed models of Otis et al. (1978) be used to estimate abundance with the data arising from each of the short-term sampling episodes. These data then can be pooled (with each animal recorded as caught if it was observed at least once during the closed population sampling) to estimate survival based on the CJS estimators. Recruitment can be estimated using the abundance estimates from the closed models and survival estimates from the open models (Pollock 1982).

The robust design consists of k primary sampling occasions, between which the population is likely to be open to gains and losses. At each primary sampling occasion j, a short-term study is conducted, with the population sampled over l_j secondary sampling periods during which it is assumed to be closed, although this assumption can be relaxed (Schwarz and Stobo 1997; Kendall and Bjorkland 2001). As an example, a small mammal population might be trapped for five consecutive days every two months, in which case, $l_j = l = 5$.

Capture–recapture data from the robust design can be summarized as number of animals exhibiting each capture history, where a capture history contains information about both secondary and primary periods. For example, a study with $k = 4$ primary sampling periods and $l = 5$ secondary sampling periods within each primary period would produce a capture history with 20 columns. A history for a particular animal might be

$$01101 \; 00000 \; 00100 \; 10111$$

consisting of four groups of five capture values. The first group of five numbers gives the capture history over the five secondary periods of primary period 1, showing that the animal was captured on secondary occasions 2, 3, and 5 of primary period 1. The second group of numbers indicates that the animal was not captured at all during primary period 2. In primary period 3 it was captured on the third secondary occasion, and in primary period 4 it was captured on secondary occasions 1, 3, 4, and 5.

Under Pollock's (1981a, 1982) original robust design, abundance was estimated with closed models using secondary capture history data, survival rates were estimated using standard open models with capture-history data reflecting captures in each primary period, and the number of new recruits was estimated using the closed-model abundance estimates and open-model survival estimates. Thus, the modeling proceeds via independent selection of an open model that incorporates survival and capture probabilities for the primary periods, and a closed model that incorporates abundances and capture probabilities for the secondary periods.

The likelihood-based approach to the robust design (Kendall et al. 1995, 1997) differs from the ad hoc approach in that a full likelihood is described for data from both secondary and primary periods. The full likelihoods are written as products of components corresponding to the two types of data, with mathematical relationships among the capture parameters of the components. Define the capture probability parameters

p_{jg} = probability that an animal is captured in secondary period g of primary period j; and

p_j^* = probability that an animal is caught on at least 1 secondary period of primary period j.

These parameters, p_{jg}, p_j^*, are related through the following expression:

$$p_j^* = 1 - \prod_{g=1}^{l_j}(1 - p_{jg}) \tag{5.11}$$

Thus, an animal must be missed (not caught) in each of the secondary periods of primary period j to be missed in primary period j.

Consideration of the number of possible ways of modeling survival in the open-model framework and modeling capture probability in both the closed and open frameworks leads to the conclusion that the number of possible robust design models is large (Kendall et al. 1995). At one time, the inability to obtain maximum likelihood estimates for closed-population models that incorporated heterogeneous capture probabilities removed this class of closed models from consideration for joint likelihoods. However, the finite mixture models of Norris and Pollock (1996b) and Pledger (2000) now permit modeling of heterogeneity in a likelihood framework, so virtually any sort of closed model can be included in a joint likelihood. Reduced-parameter models, multiple-group models, multiple-age models, models with capture-history dependence, reverse-time models, time-specific and individual covariate models, and multistate models can all be used to develop joint likelihoods for robust design data (e.g., Kendall and Nichols 1995; Kendall et al. 1997; Nichols and Coffman 1999; Nichols et al. 2000; Lindberg et al. 2001; Kendall and Bjorkland 2001).

Assumptions underlying joint likelihood models for robust design data basically represent a union of the assumptions underlying the closed and open models that are combined. The important exception involves the capture probability parameters that are shared by the two components of the likelihoods. In the presence of temporary emigration, the relationship of equation 5.11 does not hold, as the complement of p_j^* includes temporary emigration, whereas the complements of the p_{jg} reflect only the conditional probability of being captured, given presence in the sampled area (see below). In this situation, the models based on equation 5.11 will yield biased estimates, and appropriate models must incorporate the possibility of temporary emigration. Such models have been developed by Kendall and Nichols (1995), Kendall et al. (1997), Schwarz and Stobo (1997), and Kendall and Bjorkland (2001).

Temporary Emigration

Kendall et al. (1997) introduced random and Markovian models for temporary emigration, both of which are based on the concept of a superpopulation of N_j^0 animals. These animals are associated with the area sampled at period j, in the sense that they have some nonnegligible probability of being in the area exposed to sampling efforts during period j. Some number N_j of these animals are actually in the area and therefore are available for possible capture with probability p_j^*. The models of Kendall et al. (1997) assume that the population is closed to gains and losses (including temporary emigration) over the secondary periods of primary period j, but this assumption can be relaxed if necessary (Schwarz and Stobo 1997; Kendall and Bjorkland 2001).

Define a new capture probability parameter associated with the superpopulation:

p_j^0 = the probability that a member of the superpopulation at primary period j (one of the N_j^0 animals in the superpopulation) is captured during primary period j.

The capture probability parameter p_j^* under the robust design now corresponds to the probability that an animal exposed to sampling efforts at j (one of the N_j animals) is captured during j. This capture probability can thus be viewed as conditional on presence in the sampled area. The survival rate, ϕ_j, is redefined to reflect the probability that a member of the superpopulation at time j is still alive and a member of the superpopulation at time $j + 1$.

The model for random temporary emigration (Kendall et al. 1997; Burnham 1993) requires parameters η_j representing the probability that a member of the superpopulation at period j is not in the area exposed to sampling efforts during j (i.e., is a temporary emigrant). Thus, the expected number of animals in the area exposed to sampling efforts can be written as

$$E(N_j \mid N_j^0) = (1 - \eta_j)N_j^0$$

It is also possible to specify the relationship between the capture probabilities for animals that are exposed to sampling efforts at j (p_j^*), and for those in the entire superpopulation, regardless of whether or not they are exposed to sampling efforts at j (p_j^0). This is

$$p_j^0 = (1 - \eta_j)p_j^* \tag{5.12}$$

Equation 5.12 simply specifies that for a member of the superpopulation to be caught at any period j, it must be in the area exposed to sampling efforts and then be captured.

The joint likelihood models that incorporate random temporary emigration model the secondary period data in the usual way with p_j^* written as a function of capture probabilities for closed-population models. These closed-population models permit estimation of capture probability conditional on presence in the sampling area and exposure to sampling efforts. In the open-model portion of the likelihood, however, p_j^0 is substituted for the capture probability parameters, reflecting the possibility that the animal may not be in the area exposed to capture efforts. Such modeling permits direct estimation of conditional capture probability p_j^* and temporary emigration probability η_j.

Kendall et al. (1997) also developed a class of more general models, in which the probability of being a temporary emigrant at primary period j depends on whether or not the animal was a temporary emigrant at time $j - 1$. Specifically, let η_j' denote the probability that a temporary emigrant at primary period $j - 1$ (i.e., was included in $N_{j-1}^0 - N_{j-1}$) is also a temporary emigrant at time j. Let η_j'' denote the probability that a nonemigrant at $j - 1$ is a temporary emigrant at j. Temporary emigration is thus modeled as a first-order Markov process.

To illustrate the Markovian model of temporary emigration, consider the probability associated with primary-period capture history 01010, which is

Pr(01010 | release at period 2)
$$= \phi_2\phi_3[\eta_3''(1 - \eta_4') + (1 - \eta_3'')(1 - p_3^*)(1 - \eta_4'')]p_4^*\{1 - [\phi_4 p_5^*(1 - \eta_5'')]\}$$

The above expression includes two possibilities, the probabilities for which are added together inside the first set of brackets. The first possibility is that the animal released at period 2 was a temporary emigrant at period 3. The second possibility is that the animal was not a temporary emigrant at period 3, but was simply not caught then. These two possibilities require two different temporary emigration parameters for period 4, reflecting the different emigration status at period 3.

Recruitment Components

Capture–recapture modeling for open populations provides estimates of gains to, and losses from, the sampled population. Sometimes it is of interest to decompose rates of gain into components associated with immigration and in situ reproduction. This separation is possible with the robust design using an approach described by Nichols and Pollock (1990); see also Pollock et al. (1990). Similarly, a reverse-time approach can be used to directly estimate the proportional contributions of immigration and in situ reproduction to population growth (Nichols et al. 2000).

Estimation

Robust design models with no temporary emigration and with both random and Markovian temporary emigration yield product-multinomial likelihoods. Similarly, multiple-age reverse-time modeling is based on a multinomial likelihood. Maximum likelihood estimation under most of these models is accomplished using programs MARK (White and Burnham 1999), RDSURVIV (Kendall and Hines 1999), and ORDSURVIV (Kendall and Bjorkland 2001).

5.7 Discussion

The number of capture–recapture models developed for open animal populations has increased greatly in the last two decades. This increase was made possible by high-speed computers that are able to maximize likelihood functions numerically. Between the mid-1960s and the mid-1980s, the only models available to biologists were those for which closed-form estimators had been derived (Cormack 1964; Jolly 1965; Seber 1965; Robson 1969; Pollock 1975, 1981b; Brownie and Robson 1983).

During the mid-1980s, open models requiring iterative computer solutions were developed with accompanying software (Cormack 1981; Sandland and Kirkwood 1981; Jolly 1982; White 1983; Conroy and Williams 1984; Clobert et al. 1985; Crosbie and Manly 1985; Brownie et al. 1986; Burnham et al. 1987). For the last 15–20 years, flexible software such as SURVIV (White 1983), SURGE (Pradel et al. 1990), and now MARK

(White and Burnham 1999) has permitted quantitative biologists to participate in the modeling process. This greater involvement of biologists in model development has led to a proliferation of models tailored to a variety of organisms and specific sampling situations.

The net result of this recent increase in development of open models has been an interesting shift of burden from the statistician to the field biologist. During the 1960s and 1970s, the rate of model development was limited by statisticians, as the burden fell to this group to develop the models and estimators needed to deal with various biological and sampling situations. However, over the last 15 years, statisticians and quantitative biologists have responded to this challenge and developed models and estimators for extremely complex biological and sampling scenarios. The burden has thus shifted to the biologist to conduct the appropriate experiments and collect the data needed to support these often-complicated and parameter-rich models of biological phenomena.

Finally, we note that an important future direction for capture–recapture modeling for open populations involves combination of capture–recapture data on sample areas with other data sources. Some joint models have been developed already (Freeman et al. 1992; Burnham 1993; Barker 1997, 1999; Powell et al. 2000, Besbeas et al. 2002, 2003) and others have been proposed (e.g., Bennetts et al. 2001; Nichols and Hines 2002). Such joint models offer the possibility of integrating seemingly disparate methods of studying and monitoring animal populations by bringing all relevant data to bear on specific inference problems.

5.8 Chapter Summary

- Two types of model are defined for open populations. Cormack-Jolly-Seber (CJS) models are conditional on the number of animals released on each sampling occasion, whereas Jolly-Seber (JS) models use information on the captures of unmarked animals and are therefore not conditional on the captured animals. The CJS models allow for the estimation of capture and survival probabilities, while the JS models also allow the estimation of abundance and other related parameters.
- The probabilistic basis for CJS models is explained, and the assumptions are noted. Reduced-parameter versions of the models are reviewed, with capture or survival probabilities assumed to be constant over time. The models can be extended through the use of covariates to account for variation in parameters over time, variation in parameters among groups of individuals (e.g., males and females), and variation in parameters associated with individuals.

- Model selection and testing for goodness-of-fit are discussed.
- An example involving the trapping of meadow voles is used to illustrate the various methods.
- Models that allow parameters to vary with the known age of animals are discussed, including models for animals first released at age zero, and age-specific breeding models.
- Reverse-time models are reviewed as a means of estimating parameters associated with the recruitment process.
- The unconditional JS types of models are reviewed, including the use of Horvitz-Thompson estimators of population size, the superpopulation model of Crosbie and Manly, and the temporal symmetry approach for the estimation of a population growth rate. Methods are illustrated using the meadow voles example.
- The robust capture–recapture sampling design is described, involving several periods of intensive sampling when the population size can be assumed to be constant, separated by longer periods during which the population size is expected to change. Estimation methods specific to this design are reviewed, including the allowance for temporary emigration.
- The general development of models for open-population capture–recapture data in the past is discussed, and likely future developments are anticipated.

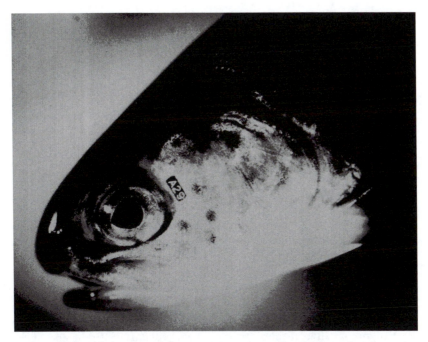

Figure 5.3. Visible implant alpha-numeric tag in steelhead (*Oncorhynchus mykiss*) smolt. (Courtesy of Northwest Marine Technology, Inc.)

Six

Tag-recovery Models

JOHN M. HOENIG, KENNETH H. POLLOCK,
AND WILLIAM HEARN

6.1 Introduction

Modern tagging models for estimating mortality rates of exploited populations derive from the work of Seber (1970), Brownie (1973), Youngs and Robson (1975), and Brownie et al. (1985). These models pertain to the case where tagged animals are killed when they are recaptured and there is no direct information on animals that die of natural causes (such as empty shells of mollusks). The authors cited above concentrated on the estimation of the annual survival rate, S, which is the probability that an animal alive at the start of the year will survive to the end of the year. These models commonly are called Brownie models. Rates of harvesting can also be derived from Brownie models under special circumstances. However, a class of models known as instantaneous rates models has also been developed specifically for estimating fishing or hunting mortality and natural mortality (the latter being the rate of mortality due to all causes other than harvesting).

Modern tagging models are based on the following logic. If two tagged cohorts of animals that are completely vulnerable to hunting or fishing are released exactly one year apart in time, then the fraction of the tags recovered in a unit of time (any time after the release of the second cohort) would be the same if the two cohorts had experienced the same cumulative mortality. However, the first cohort has been at liberty one year longer than the second, and has consequently experienced more mortality. Therefore, the rate of return of tags from the first cohort should be lower than from the second, reflecting the amount of mortality in the first year.

The structure of a tagging study is to have R_i animals tagged at the start of year i, for $i = 1, 2, \ldots, I$. There are then r_{ij} recaptures during year j from the cohort released in year i, with $j = i, i + 1, \ldots, J$ and $I \leq J$, where, the term "cohort" refers to a batch of similar (e.g., similarly sized) animals tagged and released at essentially the same time. The recapture data can be displayed in an upper triangular matrix (table 6.1).

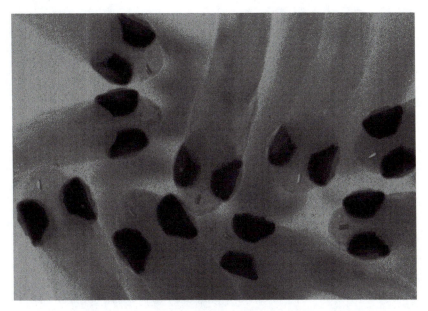

Figure 6.1. Half-length "Coded Wire Tag" implanted into 37-mm-long pink salmon fry (*Oncorhynchus gorbuscha.*). (Courtesy of Northwest Marine Technology, Inc.)

It is also possible to construct a table of expected recaptures corresponding to the observed recaptures. Let S_j be the survival rate in year j, and f_j (the tag-recovery rate) be the probability a tagged animal is recaptured and the tag reported to the investigator in year j. Then the expected number of tag returns in year j, from animals tagged in year i, is given by (table 6.2)

$$E(r_{ij}) = R_i f_j$$

if $i = j$, and

$$E(r_{ij}) = R_i \prod_{h=i}^{j-1} S_h f_j \tag{6.1}$$

if $i < j$. In the absence of further information, the parameter f_j is uninformative because the fraction of tags returned in a year depends on the amount of hunting or fishing effort, the reporting rate of tags, the extent to which tagged animals survive the tagging operation, and the extent to which tags are retained by the animals. In some cases, tag-induced mortality and tag loss can be considerable.

TABLE 6.1
Lake trout tag-return data from Hoenig et al. (1998a) showing the general
structure of a tag-recovery matrix

Year tagged	Number tagged	Recaptures in year				
		1	2	3	4	5
1960	1,048	48	55	10	11	5
1961	844	—	49	39	26	9
1962	989	—	—	36	56	15
1963	971	—	—	—	49	31
1964	863	—	—	—	—	32

Note. These data were modified from those presented by Youngs and Robson (1975) to
simulate a case of nonmixing of newly tagged animals where exploitation rate for newly
tagged animals is 2/3 that of previously tagged animals.

The logic outlined above can be formalized by considering a moment
estimator of the survival rate S_i. For two cohorts i and i' released one
year apart ($i' = i + 1$), a moment estimate is found as the solution of

$$\frac{r_{ij}}{r_{i'j}} = \frac{R_i\,\Pi_{h=i}^{j-1}\,S_h f_j}{R_{i'}\,\Pi_{h=i'}^{j-1}\,S_h f_j} = \frac{R_i S_i}{R_{i'}}$$

Thus, the estimate is

$$\hat{S}_i = \frac{R_{i'} r_{ij}}{R_i r_{i'j}}$$

For example, if 844 animals are released with tags at the start of 1961
and 989 animals are released at the start of 1962, and if the recoveries in
year 4 of the study (1963) are $r_{1961,4} = 26$ while $r_{1962,4} = 56$ (table 6.1),
then an estimate of survival in 1961 is

$$\hat{S}_{1961} = \frac{989 \times 26}{844 \times 56} = 0.544$$

Also, note that the number of recaptures from animals tagged in 1961 and
recaptured in 1961 is 49 (table 6.1). The expected number for this cell is
$R_{1961} f_{1961}$ (table 6.2). Equating the observed and expected recaptures
gives $49 = 844\,f_{1961}$. This gives us an estimate of $49/844 = 0.058$ for
f_{1961}. This parameter is discussed further in section 6.3.

In practice, parameters are estimated from all of the data by using the
method of maximum likelihood. The recaptures from a cohort over time are
assumed to constitute a random sample from a multinomial distribution.

TABLE 6.2

Cell probabilities π_{ij} for a general Brownie tagging model with $I = 3$ years of tagging and $J = 4$ years of recaptures

Year	Number tagged	Cell probability for year				Probability never recaptured
		1	2	3	4	
1	R_1	f_1	$S_1 f_2$	$S_1 S_2 f_3$	$S_1 S_2 S_3 f_4$	$1 - f_1 - S_1 f_2 - S_1 S_2 f_3 - S_1 S_2 S_3 f_4$
2	R_2	—	f_2	$S_2 f_3$	$S_2 S_3 f_4$	$1 - f_2 - S_2 f_3 - S_2 S_3 f_4$
3	R_3	—	—	f_3	$S_3 f_4$	$1 - f_3 - S_3 f_4$

Note. The expected number of recaptures $E(r_{ij})$ is found by multiplying the cell probability by the number of animals tagged and released in the cohort, R_i.

That is, an animal tagged in year i can be recaptured in year $i, i + 1, \ldots,$ J, or not be recaptured at all. Thus, the recaptures from cohort i constitute a sample from a multinomial distribution with $J - i + 2$ categories. The likelihood function for tagged cohort i is then proportional to

$$L = \left(1 - \sum_{j=i}^{J} \pi_{ij} \right)^{R_i - \sum_{j=i}^{J} r_{ij}} \cdot \prod_{j=i}^{J} \pi_{ij}^{r_{ij}} \qquad (6.2)$$

where the cell probabilities, π_{ij}, are as in table 6.2. The factor before the product refers to the tagged animals never recovered, and the likelihood for all of the cohorts is simply the product of the likelihoods for each cohort, which are considered independent.

The model represented in table 6.2 is very general in that it allows for year-specific values of both the survival rate and the tag-recovery rate. It may be of interest to construct more restricted models. For example, if fishing or hunting effort has been constant over time, it may be of interest to fit a model with a single survival rate for all years. Similarly, if fishing or hunting effort or tactics changed during the course of the study, it may be of interest to fit a model with one survival rate for the period before the change and another survival rate for the period after. The likelihood is still constructed as in (6.2) but the cell probabilities are defined in terms of the restricted set of parameters of interest. From the above, it is clear that it is possible to fit many related models to a single data set. Modern computer packages make this easy with their computational power and interactive interfaces. A key feature has been the use of the Akaike Information Criterion (AIC) for model selection (Burnham and Anderson 1998) to arrive at a parsimonious model, as discussed in chapter 1.

Important things to note are that (1) the Brownie models enable the estimation of the survival rate (or total mortality rate), but not the components of mortality (harvest and natural mortality); (2) to obtain one estimate of the survival rate it is necessary to tag two cohorts of animals, so that the first estimate of survival can be obtained at the end of the second year of study; and (3) it is not possible to estimate the survival rate in and after the most recent tagging year with this model.

6.2 Assumptions of Brownie Models

The assumptions of Brownie type models are (Pollock et al. 1991, 2001) that

1. the tagged sample is representative of the population being studied;
2. there is no tag loss or, if tag loss occurs, a constant fraction of the tags from each cohort is lost, and all tag loss occurs immediately after tagging;
3. the time of recapture of each tagged animal is reported correctly, although sometimes tags can be returned several years after the animals are recaptured;
4. all tagged animals in a cohort have the same survival and recovery rates;
5. the decision made by a fisher or hunter on whether or not to return a tag does not depend on when the animal was tagged;
6. the survival rate is not affected by tagging or, if it is, the effect is restricted to a constant fraction of animals dying immediately after tagging; and
7. the fate of each tagged animal is independent of that of the other tagged animals.

There are several common problems in tagging studies. Newly tagged animals may not have the same spatial distribution as previously tagged animals, especially if animals are tagged at just a few locations, which leads to failure of assumption 1. Animals of different sizes or ages are tagged and these have different survival rates due to size or age selectivity of the harvest, which leads to failure of assumption 4. Older animals have a different spatial distribution than younger animals (due to different migration patterns or emigration), which leads to failure of the assumptions 1 and 4. Tests for these problems and possible remedies are discussed below.

6.3 Interpretation of the Tag-recovery Rate Parameter

As indicated above, the parameter f_j of the Brownie model is a composite parameter. The rate of return of tags (i.e., the fraction of surviving tagged animals that is recovered) can be modeled as $f_j = \theta_j u_j \lambda_j$, where θ_j is the probability an animal tagged in year j survives the tagging procedure and retains its tag, u_j is the probability a tagged animal alive at the start of year j will be captured in year j, and λ_j is the probability that an animal recaptured in year j will be reported. The parameter u is often referred to as the exploitation rate. If θ_j and λ_j can be estimated, then an estimate of the tag-recovery rate parameter, f_j, can be converted into an estimate of the exploitation rate, u_j (Pollock et al. 1991).

For example, in section 6.2 there was an estimate of f of 0.058 for 1961. Suppose we have determined that all animals survive the tagging procedure with the tag intact ($\theta = 1$) and that 20% of the hunters or fishers who catch a tagged animal will report the tag ($\lambda = 0.2$). Then the exploitation rate in 1961 is estimated to be $0.058/(1 \times 0.2) = 0.29$.

Estimates of θ_j can be obtained by holding newly tagged animals in enclosures and observing the proportion surviving with tags intact (e.g., see Latour et al. 2001c). One way to estimate tag-reporting rate is by using high reward tags (Pollock et al. 2001, 2002a). Suppose in a tagging study some tags have the standard reward value while others have a special, high reward. It is assumed that all the high reward tags recovered by fishers or hunters are reported. This implies that the value of the reward provides a powerful incentive to return the tag, and that high reward tags are recognized when encountered, e.g., because of a publicity campaign. If the rate of return of standard tags is one-third the rate of return of high reward tags, this implies the fraction of standard tags that is reported is 1/3 (Henny and Burnham 1976). A variety of monetary reward values can be used so that the rate of return of tags can be studied as a function of the reward value (Nichols et al. 1991). Sometimes investigators release high reward tags after a tagging program has been in existence for a number of years. This is likely to change the behavior of the fishers or hunters so the estimates of harvest rates as well as tag-reporting rates pertain to when there is a high reward tagging program in operation but not to the time prior. Pollock et al. (2001) discuss how to handle this situation.

In some cases the investigator would be better off using only high reward tags, instead of a combination of high rewards and standard rewards, even though this leads to many fewer tagged animals being released (Pollock et al. 2001). In essence, high reward tags provide much more powerful information, so the reporting rate for standard tags must be high to justify their use.

Another way to estimate reporting rate is by using hunting or fishing catch surveys (Pollock et al. 1991; Hearn et al. 1999; Pollock et al. 2002a,b). Suppose 5% of the hunting or fishing activity is observed in a hunter, creel, or port sampling survey or by onboard observers. For example, there might be 12 access points to a hunting area and 100 days of hunting; thus, there are 1200 access point-days of hunting activity of which 60 are randomly sampled. Here, sampling an access point-day means observing all hunters' kill at the access point on the specified day. Suppose further that 15 tags are recovered by the survey agents. This number of recoveries is converted into an estimate of the total number of tagged animals caught by dividing the number of tagged animals observed in the survey by the fraction of the hunt sampled. Thus, the estimated number of tagged animals caught = 15/0.05 = 300. Then, if hunters report all tags, we would expect to receive returns of 300, consisting of the 15 tags already recovered plus 285 additional tags. If only 95 tags (one-third of 285) are recovered outside of the survey, the tag reporting rate is estimated to be 95/285 = 1/3 (Pollock et al. 1991).

Similar methods using fishery observer programs were developed by Paulik (1961), Kimura (1976), and Hearn et al. (1999), who estimated the number of tags in the catch, and by Pollock et al. (2002b), who combined the estimation of reporting rate with estimation of mortality rates. Suppose there are 30,000 metric tons (mt) of fish landed in a fishery, and observers on boats observe 3000 mt of catch (i.e., one-tenth of the total catch) and recover 12 tags. Then it is estimated that 12/0.1 = 120 tagged fish were caught in the fishery, and it should be possible to recover 120 –12 = 108 more tags if the fishers reported all tagged animals caught. This method differs from the previous method in that here it is assumed that the catches are known, whereas in the previous method the fishing opportunities (the access point by day combinations) were known.

Another method of estimating tag-reporting rate involves planting tagged animals among a fisher's catch (in a way that does not alter the fisher's behavior) and noting the fraction of the tags that is recovered (Costello and Allen 1968; Green et al. 1983; Campbell et al. 1992; Hearn et al. 2003). Hearn et al. (2003) discuss planting tags in one component of a multicomponent fishery.

It is important that planted tags be placed in the catch before any humans examine the catch. This generally means that someone must plant the tagged fish before the crew processes the catch. Otherwise, the study will pertain only to the reporting rate for those steps in the processing of the catch that occur after the tags are planted. This method is unlikely to be useful in recreational hunting and fishing studies because it is almost impossible to plant tags surreptitiously soon after capture.

It is also possible to obtain an estimate of the tag-reporting rate (λ_j) from the tagging data itself (Youngs 1974; Siddeek 1989, 1991; Hoenig et al. 1998a). In general, the information about reporting rate in the tagging data is very weak. However, tagging studies can be designed to enhance the ability to infer reporting rates from tagging data (Hearn et al. 1998; Frusher and Hoenig 2001).

6.4 Functional Linkage Between the Exploitation Rate and the Survival Rate

An important point is that the exploitation rate is not linked functionally to the survival rate in the Brownie model. For example, from one year to the next, both the survival rate and the exploitation rate estimates could increase. Biologically, this is possible if natural mortality decreases over time. In fisheries studies, it is common to assume that the natural mortality rate is constant (or that it fluctuates randomly and independently of exploitation). In this case it is more efficient to use a model in which increasing exploitation rate implies a reduced survival rate. This property is found in the instantaneous rates models developed by Hoenig et al. (1998a,b) and applied by Latour et al. (2001c) and Frusher and Hoenig (2001). These are discussed below. This aspect of tagging models is controversial. Particularly in wildlife studies, investigators are often hesitant to use a linked model, but Anderson and Burnham (1976) and Nichols et al. (1984a) have suggested that a compensatory model is sometimes warranted.

6.5 Instantaneous Rate Models for Estimating Harvest and Natural Mortality

Under simple competing risks theory, in any instant of time, harvest and natural mortality operate independently and additively. The probability of surviving a given period of time of duration t is related to the instantaneous harvest (H) and natural mortality (A) rates in the period by $S = e^{-(H+A)t}$. Here, H and A have units of time^{-1} so that $(H+A)t$ is without units. Note that this relationship holds regardless of the relative timing of the harvest and natural mortality. In fisheries studies, H and A are called the instantaneous rates of fishing and natural mortality, and are denoted by F and M, respectively.

The exploitation rate, u, can be defined as the fraction of the population present at the start of the year that is harvested during the year. It is a function of the harvest and natural mortality rates. The exact functional form depends on the relative timing of the forces of mortality. If harvest and natural mortality operate throughout the year at constant intensity, in what Ricker (1975) calls a type two fishery, then the exploitation rate is

$$u = \frac{H}{H+A}(1 - e^{-(H+A)}) \tag{6.3}$$

If all the harvest mortality occurs in a restricted period of time at the beginning of the year (a type I fishery), then

$$u = 1 - e^{-H} \tag{6.4}$$

Hoenig et al. (1998a) presented an approach for modeling exploitation rate as a function of an arbitrary pattern of harvest mortality over the course of a year. However, Youngs (1976) and Hoenig et al. (1998a) found that the computed annual exploitation rate is rather insensitive to the timing of the harvest. Thus, in many cases (6.3) or (6.4) will provide an adequate approximation.

A Brownie model can be converted into an instantaneous rates model by simply replacing the S_j in the Brownie model by $\exp(-H_j - A_j)$, and replacing the f_j by $\theta \lambda u_j$, where u_j is defined by equations 6.3 and 6.4. Alternatively, the more general formulation of Hoenig et al. (1998a) can be used.

6.6 Diagnostics and Tests of Assumptions

Recently, diagnostic procedures have been developed for evaluating models and testing goodness of fit. Latour et al. (2001b) studied patterns in residuals from Brownie models and instantaneous rates models and found that certain types of failures of assumption give rise to particular patterns in the residuals (table 6.3). For example, the residuals in table 6.4 were obtained by fitting an instantaneous rates model that assumes complete mixing of newly tagged animals to a dataset that was modified to simulate newly tagged animals experiencing a lower mortality rate than previously tagged animals, as might occur if tagging took place in areas far from the main fishing or hunting grounds. The fitted model had $\theta \lambda$ fixed at 0.18. A single natural mortality rate, A, was estimated under the assumption that natural mortality is constant over all years. Year-specific harvest rates, H, are estimated. See Hoenig et al. (1998b) for a description of the other characteristics of the model. Note that this causes all residuals on the main diagonal to be negative and all residuals one cell to the right of the main diagonal to be positive.

Latour et al. (2001b) noted that the residuals from the Brownie-type model with time-varying survival and tag-recovery rates are subject to some constraints that do not apply to the instantaneous rates models. Specifically, in the time-varying parameterization of the Brownie models the residuals for the "never seen again" column must always be zero (unless constraints are imposed), the sum of the residuals in a row must be zero, the sum of the residuals in a column must be zero, and the residual

TABLE 6.3
Patterns in residuals from Brownie and instantaneous rates models caused by failures of assumption

Assumption failure	Pattern in the residuals
1. Change in tagging personnel or procedure from year to year leading to variable levels of tag induced mortality (assumption 2)	Row effects in instantaneous rates models, but not in Brownie models
2. Emigration of the oldest animals (assumption 6)	Diagonal effects, upper right corner of residuals matrix negative
3. Nonmixing of newly tagged animals (assumption 1)	Diagonal effects, may include upper right corner of residuals matrix
4. Change in tag reporting rate from year to year	No detectable effect in the residuals, which is not a problem in Brownie models, and only a problem in instantaneous rates models if the reporting rate is assumed constant
5. Change in natural mortality rate A from year to year	Not a problem for Brownie models year and causes column effect in instantaneous rates models if A is assumed constant

Note. An effect refers to a predominance of residuals of one sign.

for the $(1, 1)$ cell must be zero. If the number of years of tagging (I) equals the number of years of recapture, then the (I, I) cell residual must be zero. The instantaneous rates model is only subject to the constraint that the row sums must be zero. The additional constraints associated with the Brownie model make it difficult to visualize some patterns in the residuals. For example, suppose in year 2 of a study a defective batch of tags is used or an inexperienced tagger is employed. One might expect that the number of returns from this cohort over time would be especially low (assumption 2 is violated). With the instantaneous rates model, this is likely to result in a row of negative residuals (with the never-seen-again cell being positive). With the Brownie model, the fact that the sum of the row residuals (excluding the never seen again category) must be zero implies that there are positive residuals to balance the negative ones. Consequently, it is difficult to detect a bad batch of tags by examining Brownie model residuals. Latour et al. (2001b) suggest examining residuals from an instantaneous rate model even when one is interested in fitting a Brownie model.

TABLE 6.4

Table of residuals from the fit of a model that incorrectly assumes that newly tagged animals are fully mixed into the population to the data in table 6.1

Year tagged	Year of Recapture					Never seen again
	1	2	3	4	5	
1960	−3.9	8.0	−8.5	−7.4	−2.5	14.3
1961	—	−6.3	17.2	4.3	0.1	−15.3
1962	—	—	−7.0	13.4	−2.5	−3.9
1963	—	—	—	−9.4	7.0	2.4
1964	—	—	—	—	−2.1	2.1

Note. True exploitation rate of newly tagged animals was 2/3 that of previously tagged animals. Note that all residuals along the main diagonal are negative and all residuals in the diagonal one cell to the right of the main diagonal are positive.

Latour et al. (2001a) described procedures for testing whether newly tagged animals are fully mixed with previously tagged animals (assumption 1). This involves examining the spatial distribution of tag returns by cohort. If tagged animals are well mixed into the population, then the fraction of the recoveries in a year coming from any particular region should be the same for all tagged cohorts. This can be tested with a contingency table analysis where rows represent tagged cohorts, columns represent tag recovery locations, and the cells specify how the recoveries from a cohort are apportioned among regions.

Myers and Hoenig (1997) explored whether all animals are equally catchable (assumption 6) by examining rates of returns by the size class of animal, i.e., they estimated a selectivity curve for a fishery. The same approach can be used to study whether the rate of tag return varies by other factor such as the sex of the animal. For example, if 10% of the tagged females are recaptured, then (approximately) 10% of the tagged males should be recaptured if the assumption of equal catchability is met. If this condition is not met, the males and females can be analyzed separately to avoid biased estimates.

6.7 Preventing and Dealing with Failures of Assumptions

Latour et al. (2001c) analyzed a tagging study of juvenile red drum in a South Carolina estuary. A feature of the drum population is that the juveniles remain in the estuary for only a few years. Consequently, two or three years after tagging, a cohort will disappear due to permanent emigration. This phenomenon was evident from the pattern of negative residuals in the upper right corner of the residuals matrix. To deal with this, Latour et al. chopped off the upper right corner of the recapture matrix and added the observations to the column specifying the number of animals never seen again. Suppose, for example, animals are tagged for five years and recovered over a five-year period (as in table 6.1). Suppose also that the residuals for the (row = 1, column = 4), (1,4) and (2,5) cells are negative. Then, for row 1, the recaptures r_{15} and r_{14} are added to the number of tags never seen again, thus resulting in a tetranomial $(r_{11}, r_{12}, r_{13}$ and $R_1 - r_{11} - r_{12} - r_{13})$. Similarly, for row 2, the recaptures r_{25} are added to the never seen again category resulting in a tetranomial $(r_{22}, r_{23}, r_{24}$ and $R_2 - r_{22} - r_{23} - r_{24})$. The likelihood is constructed as before by raising the cell probability to the observed number of recaptures in the cell for every defined cell. The "chop" approach is applicable to both Brownie and instantaneous rates models. It results in a loss of precision (because data are discarded) but it can reduce bias.

Suppose males and females have different survival rates. One solution is to simply conduct two tagging studies, one on each sex. This is a legitimate approach but it is expensive and may not be the most efficient way to proceed. It may be that the female survival and recovery rates are functionally related to the male rates, or that the components of mortality are functionally related. For example, suppose that males and females have a different natural mortality rate but the same fishing (or hunting) mortality rates. The table of expected recaptures for a type 1 fishery when there are $I = 3$ years of tagging and $J = 4$ years of recaptures is shown in table 6.5. Note there are twice as many rows as years of tagging. This model has three fishing (hunting) mortality rates and two natural mortality rates. If separate models were fitted to each sex, there would be six fishing (hunting) mortality rates and two natural mortality rates. An example of this approach is provided by Frusher and Hoenig (2001).

Often, the survival rate depends on the age of the animal, which, in turn, may depend on the size of the animal. Thus, heterogeneity of survival rates is caused by variation in size of the animals tagged. Brownie et al. (1985) discuss a model that can be applied to age- or size-structured populations. Suppose animals of age 4 are tagged in year 1 and animals of age 5 are tagged in year 2. Let R_{ij} be the number of animals age i tagged in year j, S_{ij} be the survival rate of animals age i in year j, and f_{ij} be the tag recovery rate of animals age i in year j. The expected recaptures for a Brownie model for these two cohorts is shown in table 6.6. The logic of the study design is as follows. In year 1, the 4-year-olds in cohort 1 undergo the survival rate of age 4 animals in year 1. The next year the animals in cohort 1 are age 5; consequently, we tag 5-year-old animals to see the contrast between two groups that differ only in that the first group has experienced one more year of mortality than the second group. If we let r_{ij} be the number of recaptures of animals tagged at age i

TABLE 6.5
Cell probabilities π_{ij} for an instantaneous rates tagging model with $I = 3$ years of tagging, $J = 4$ years of recaptures, and $G = 2$ sexes

Year	Number tagged	Cell probability for recapture year[a]			
		1	2	3	4
Males					
1	R_{1m}	$\theta\lambda u_1$	$\exp(-\mathbf{H}_1 - A_m)\theta\lambda u_2$	$\exp(-\mathbf{H}_1 - H_2 - 2A_m)\theta\lambda u_3$	$\exp(-\mathbf{H}_1 - H_2 - H_3 - 3A_m)\theta\lambda u_4$
2	R_{2m}	—	$\theta\lambda \mathbf{u}_2$	$\exp(-\mathbf{H}_2 - A_m)\theta\lambda u_3$	$\exp(-\mathbf{H}_2 - H_3 - 2A_m)\theta\lambda u_4$
3	R_{3m}	—	—	$\theta\mathbf{u}_3$	$\exp(-\mathbf{H}_3 - A_m)\theta\lambda u_4$
Females					
1	R_{1f}	$\theta\lambda u_1$	$\exp(-\mathbf{H}_1 - A_f)\theta\lambda u_2$	$\exp(-\mathbf{H}_1 - H_2 - 2A_f)\theta\lambda u_3$	$\exp(-\mathbf{H}_1 - H_2 - H_3 - 3A_f)\theta\lambda u_4$
2	R_{2f}	—	$\theta\lambda \mathbf{u}_2$	$\exp(-\mathbf{H}_2 - A_f)\theta\lambda u_3$	$\exp(-\mathbf{H}_2 - H_3 - 2A_f)\theta\lambda u_4$
3	R_{3f}	—	—	$\theta\mathbf{u}_3$	$\exp(-\mathbf{H}_3 - A_f)\theta\lambda u_4$

[a] To incorporate incomplete mixing of new animals into this model, replace parameters in bold typeface with u^* and H^*, respectively (e.g., for years 1 through 3 there will be abnormal harvest rates H^* to estimate and for years 2, 3, and 4 there will be normal harvest rates, H (see section 6.7)).

Note. The expected number of recaptures $E(r_{ijs})$ for sex g is found by multiplying the cell probability by the number of animals tagged in cohort R_{is}. Males and females are assumed to have the same harvest rates but different natural mortality rates. The exploitation rate, u_j, is thus $u_j = 1 - \exp(-H_j)$. The column pertaining to animals never seen again is omitted for convenience.

TABLE 6.6

Expected recoveries for an age-structured Brownie model with $I = 2$ years of tagging and $J = 4$ years of recaptures

Year tagged	Number tagged	Expected recaptures in year			
		1	2	3	4
1	R_{41}	$R_{41}f_{41}$	$R_{41}S_{41}f_{52}$	$R_{41}S_{41}S_{52}f_{63}$	$R_{41}S_{41}S_{52}S_{63}f_{74}$
2	R_{52}	—	$R_{52}f_{52}$	$R_{52}S_{52}f_{63}$	$R_{52}S_{52}S_{63}f_{74}$

Note. R_{ij} is the number of animals of age i tagged in year j; f_{ij} and S_{ij} are the tag recovery rate and annual survival rate, respectively, for animals (currently) age i in year j.

and recovered in year j, the ratio of recaptures from the two cohorts in any recovery year (after the first year) can be used to estimate S_{41}. That is, r_{42}/r_{52} estimates

$$\frac{R_{41}S_{41}f_{52}}{R_{52}f_{52}} = \frac{R_{41}}{R_{52}}S_{41}$$

and r_{53}/r_{63} estimates

$$\frac{R_{41}S_{41}S_{52}f_{63}}{R_{52}S_{52}f_{63}} = \frac{R_{41}}{R_{52}}S_{41}$$

This approach can also be used with instantaneous rates models. It essentially requires one to do a separate tagging study for each age group in each year. Again, increased efficiency can sometimes be obtained by modeling the relationships between the age groups. For example, if the fishing gear and fishing tactics remain constant over time, it may be reasonable to assume that the fishing mortality for one age group is a constant fraction of that of another age group (i.e., there is age selectivity). To our knowledge, this has not applied to tag-return models though it is commonly done in the analysis of catch at age data in fisheries assessments.

Perhaps the most critical assumption of a tagging study is that each cohort of tagged animals is representative of the population under study. This implies that all cohorts mix thoroughly with the population so that all cohorts experience the same exploitation rate. However, particularly when animals are tagged and released at just a few locations, the tagged animals may not have time to mix. Ideally, animals should be released at many sites throughout the geographic range of the population, and the number tagged at any site should be proportional to the local abundance at the site. Local abundance is often judged by the catch rate of animals. Using catch rate to assess abundance, however, is reliable only if catch rates are uniform across sampling locations. Assuming that the catch efficiency is uniform among sample locations, tagging 30% of the animals caught in each sample would be appropriate but tagging 30 animals from each sample would not (unless the animals were uniformly distributed over space). If animals can be tagged and released far in advance of the harvest, then the tagged animals will have more time to mix throughout the population. Questions regarding catch efficiency at each location, however, still must be addressed because differences will make it difficult to assess whether representation of tags is geographically proportional.

Hoenig et al. (1998b) presented an instantaneous rates model to handle the situation where newly tagged animals are not thoroughly mixed throughout the population. However, it requires that previously tagged animals (at liberty for at least a year) be thoroughly mixed into the population. The table of expected recoveries for a study of a type 1 (pulse) fishery or hunt with $I = 3$ years of tagging and $J = 4$ years of recoveries is shown in table 6.5. Notice that in each year of tagging the newly tagged animals experience a harvest mortality rate of H_j^* (the asterisk denotes an abnormal mortality rate, i.e., a rate not experienced by the population as a whole). In contrast, in any recovery year, animals at liberty for more than a year experience the normal harvest mortality rate (assuming they are well mixed into the population at large). Notice also that in each year after the year of tagging, the number of animals available to be caught by the harvesters depends on what happened in the previous year(s). Thus, the cumulative survival rates specified in all cells above the main diagonal have a component of mortality that was abnormal. With this model, it is necessary to estimate I abnormal harvest mortality rates (one for each year of tagging), and $J - I$ normal harvest mortality rates. It is not possible to estimate the normal harvest mortality rate, H_1, in the first year of the study because all of the animals in the first year experience the abnormal harvest mortality rate.

The necessity of estimating the abnormal harvest mortality rates results in a significant loss in precision. Thus, it may well be worth the extra costs associated with dispersing the tagging effort spatially so that a model based on the assumption of proper mixing can be used instead of a model that allows for nonmixing in the year of tagging. Hoenig et al. (1998b) showed that, if a model based on assumed mixing is fitted when a nonmixing model is appropriate, the parameter estimates can be severely biased. However, the estimated standard errors will tend to be small and the analyst may be fooled into thinking that the mortality rates are known precisely when in fact the true values are far from the estimated ones.

To illustrate the importance of accounting for nonmixing, we consider data on lake trout from Youngs and Robson (1975) and modified by Hoenig et al. (1998b). The original data appeared consistent with the assumption of complete mixing, and this was not unexpected given the study design. However, Hoenig et al. modified the number of recaptures along the main diagonal by multiplying them by 2/3 to simulate newly tagged animals having an exploitation rate 2/3 that of previously tagged animals. The number of recaptures in subsequent years were adjusted upward to account for the higher survival rate in the year of tagging. The resulting data are shown in table 6.1.

The fit of the model that assumes full mixing is extremely poor (chi-squared = 34.45, with 9 df, $p < 0.0001$). The residuals (table 6.4) are all negative on the main diagonal and positive immediately to the right of the main diagonal. On the other hand, the nonmixing model fits well (chi-squared = 7.62, with 5 df, $p = 0.18$) and there is no pattern to the residuals. For the model assuming complete mixing, the negative of the log likelihood is 1985.3 and six parameters are being estimated. Consequently, the AIC is $2 \times 1985.3 + 2 \times 6 = 3982.6$. For the model assuming nonmixing of newly tagged animals, the log likelihood is 1972.7, the number of parameters is ten, and the AIC is 3965.4. Therefore, the model with nonmixing has the lower AIC and would be selected over the model with complete mixing.

Notice that the estimates from the model assuming complete mixing are very different from those derived from the model allowing for non-mixing in the year of tagging (table 6.7). Also, the standard errors for the nonmixed model are considerably higher than the corresponding standard errors in the fully mixed model. In essence, inappropriate use of the fully mixed model results in biased estimates with misleading standard errors so that they are precisely wrong. The estimates from the non-mixing model are essentially unbiased, and have standard errors that are appropriate and reflect the penalty one pays for having to estimate additional parameters.

Hoenig et al. (1998b) also showed that it may be possible to fit a model that allows for nonmixing to be present during only a part of the year of tagging. This allows more of the tagging data to be used for estimating the normal mortality rates.

Sometimes tagging models have the correct parameter structure but have more variability than predicted under a multinomial model. This would be reflected in reasonable residual patterns but with goodness of fit tests that are highly significant. It has now become standard practice to use variance inflation factors in such situations to adjust the estimated standard errors upward (see chapter 1). Of course, whenever possible, an effort should be made to reduce this problem by tagging few animals at a large number of sites rather than a lot of animals at a few sites.

TABLE 6.7
Model fits to the data in table 6.1

	Fully mixed model			Nonmixed model	
	Estimate	Standard error		Estimate	Standard error
			H_1^*	0.36	0.06
			H_2^*	0.38	0.06
			H_3^*	0.25	0.05
			H_4^*	0.37	0.07
			H_5^*	0.25	0.06
H_1	0.33	0.05			
H_2	0.47	0.05	H_2	0.82	0.14
H_3	0.29	0.04	H_3	0.46	0.08
H_4	0.42	0.06	H_4	0.69	0.16
H_5	0.26	0.05	H_5	0.41	0.13
A	0.05	0.05	A	0.12	0.05

Note. The product $\theta\lambda$ was set to 0.18 based on external information and A^* indicates an abnormal harvest mortality, i.e., mortality experienced by newly tagged animals. Note that H_1 cannot be estimated in the nonmixed model because in the first year of tagging there are no previously tagged animals and hence there are no tagged animals to experience the normal harvest mortality.

6.8 Chapter Summary

- The logical basis of tagging models (with animals recovered dead) is explained.
- The assumptions required for these models are briefly discussed.
- The interpretation of the tag-recovery parameter in terms of survival, capture, and recovery probabilities is discussed.
- The relationship between the exploitation rate and the survival rate is discussed.
- Models for the estimation of instantaneous harvest and natural mortality rates are discussed.
- Methods for assessing whether assumptions hold and tests for the goodness of fit of models are reviewed, as are approaches for preventing and dealing with failures of assumptions.

Figure 6.2. Bristle-thighed curlew (*Numenius tahitiensis*), marked with Darvic leg bands and leg flag, Seward Penninsula, Alaska, 1990. (Photo by Robert Gill)

Seven_____

Joint Modeling of Tag-recovery
and Live-resighting Data

RICHARD J. BARKER

7.1 Introduction

Between 1985 and 1990, 6160 paradise shelducks (*Tadorna variegata*) were banded in a study carried out in the Wanganui Region, New Zealand. Molting shelducks were trapped in January and birds checked for tags. Marked birds had their number recorded and were then released; unmarked birds were marked and released. Between marking occasions birds were also reported shot by hunters, so the reencounter data represents a mix of live recaptures, for which the models of chapters 3 and 5 are suitable, and dead-recovery data, for which the models of chapter 6 are suitable. Because birds were reencountered in two ways, a fully efficient analysis requires new models that simultaneously model these two reencounter processes.

The Cormack-Jolly-Seber (CJS) model (chapter 3) and tag-return models (chapter 6) have dominated the analysis of mark–recapture data from open populations. In the CJS model the data are from a sequence of samples (referred to as the capture occasions) consisting of marked and unmarked animals, where captures and recaptures are obtained over a short period of time during which the population is assumed closed. These capture occasions punctuate longer periods during which the population is subject to births and deaths. In the tag-return models, by contrast, the capture occasions serve only to provide a succession of releases of marked animals and it is the returns from dead animals between release occasions that are modeled. The use of information from encounters of marked animals during the open interval between capture occasions is a key difference between the tag-return models and the live-recovery models.

As in the paradise shelduck study, animals marked and released during early sample occasions are often recaptured during subsequent sample occasions. Live recaptures in band recovery studies have traditionally been ignored, partly because of the technical difficulties in simultaneously modeling the two types of data, and because live recaptures during

Figure 7.1. Tusk-banded Pacific walrus (*Odobenus rosmarus*), Round Island, Alaska, 1982. (Photo by Steven C. Amstrup)

trapping have been uncommon. Nevertheless, in some studies, live recaptures may be substantial, particularly for nonmigratory species. Similarly, during a study based on live recaptures of marked animals information may also be available from recoveries of dead animals. Supplementing the live-recapture information with recovery data should lead to improved estimation.

In a study where several kinds of information are available, ignoring one of the sources so that a simple model, such as CJS or tag-return model, can be used means that parameter estimates will be less precise than they otherwise could be (Barker and Kavalieris 2001). A joint model contains extra parameters that are needed to model the data. When these extra parameters have a biological relevance, and when data are sufficient to allow meaningful estimation and interpretation, their inclusion will mean that the study is more informative. In short, use of a joint model may provide better parameter estimates and more informative biological interpretations without the need for more animals to be marked.

Animal Movement

Although both the CJS and tag-recovery models have parameters for survival, it has long been recognized that the two survival probabilities are fundamentally different. The difference between these probabilities derives from their treatment of emigration.

In mark–recapture studies it is often difficult to obtain a random sample of animals from the study population. More commonly, a local population is sampled and marked. In the paradise shelduck study a bird was only at risk of live recapture if it chose to molt at one of the locations where birds were captured and banded. Once banded, a bird could appear in a later live-recapture sample only if (1) it survived and remained in the local population being sampled, or (2) it survived, left the local population, and then returned. Jolly (1965) and Seber (1965) recognized this and insisted that any emigration be permanent. That is, animals may leave the study area but they may not return. The consequence is that the Jolly-Seber survival parameter, usually denoted ϕ_j, represents the joint probability that the animal remains at risk of capture and does not die between samples j and $j+1$. In contrast, in tag-return models it is usually assumed that animals are exposed to the tag-recovery process (i.e., death and reporting) throughout their range. That is, there is no emigration from the study population. In this case there is no site fidelity component to the survival probability.

The assumption of permanent emigration in mark–recapture models is one that has been given relatively little attention. However, assumptions about movement should be based on known biology rather than mathematical convenience. For example, it may be unreasonable to insist that an animal that has moved off the study area cannot later return. In these studies it is important that the permanent emigration assumption be relaxed.

Burnham (1993) introduced random emigration as an alternative to the permanent emigration assumption. Under random emigration a marked animal may leave and reenter the study population, but the probability that it is at risk of capture at occasion j is the same for all animals. A generalization is Markovian temporary emigration where the probability an animal is at risk of capture at occasion j depends on whether or not it was at risk of capture at occasion $j-1$ (Kendall et al. 1997). Permanent emigration represents an extreme case of Markov emigration in which the probability that an animal is at risk of capture at occasion j is zero for an animal not at risk capture at occasion $j-1$.

7.2 Data Structure

To represent the different types of information in this study it is necessary to generalize the encounter history introduced in section 1.3. The format of program MARK (White and Burnham 1999; White et al. 2001) can be used to represent data from a study where animals can be reencountered by a live recapture, a live resighting, or a dead recovery. These events are summarized using a pair of indicator variables (L_j, D_j), with one pair for each of the j sampling periods. If at capture occasion j the animal is recaptured then L_j is assigned the value 1. Otherwise, it is assigned the value 0. If between capture occasion j and capture occasion $j + 1$ the animal is found dead and the tag is reported, then D_j is assigned the value 1. If instead the animal is resighted alive between capture occasions j and $j + 1$, then D_j is assigned the value 2. If the animal is not resighted or not recovered dead, between capture occasions j and $j + 1$, then D_j is assigned the value 0. Sometimes, an animal will be resighted alive and then be reported dead, with both events occurring in the interval $j, j + 1$. In such cases the indicator variable D_j is assigned the value 1, indicating that the animal was found dead and the earlier resighting is ignored.

The encounter histories for the paradise shelduck study are given in table 7.1. For example, the 76 animals with the encounter history 1100000000 were marked and released in January 1986 and then reported dead during the 1986 hunting season, which took place between May and August. The 16 animals with the history 1011000000 were marked and released in January 1986, recaptured in January 1987, and then reported dead during the 1987 hunting season.

The parameters of the joint live-recapture and dead-recovery model are S_j, the true survival probability between sample occasions j and $j + 1$; F_j, the probability that an animal in the local study population does not leave the study population between samples j and $j + 1$; F_j', the probability that an animal not in the local study population at the time of sample j is in the local study population at the time of sample $j + 1$; r_j, the probability that an animal that dies between samples j and $j + 1$ is found and its band is reported and p_j, which, as with capture–recapture models, is the probability of capture at occasion j.

7.3 Simple Models

Anderson and Sterling (1974) carried out an analysis of banding data from pintail ducks (*Anas acuta*) by modeling the live-recapture data separately from the tag-return data. Under the assumption that emigration is permanent, the Jolly-Seber survival estimator $\hat{\phi}_j$ provides an estimate of $F_j S_j$. Here, S_j is the probability an animal alive at sampling occasion j is still alive at sampling occasion $j + 1$, and F_j is the probability that an animal at risk of capture at j is still at risk of capture at $j + 1$ (i.e., it has not permanently emigrated). The tag-return model provides an estimate of

TABLE 7.1

Encounter histories in program MARK format for the joint analysis of live-recapture and dead-recovery data from a study of paradise shelduck banded in the Wanganui Region, New Zealand between 1986 and 1990

Encounter history	Number of animals	Encounter history	Number of animals
1100000000	76	0010101010	4
1011000000	16	0010101000	29
1010101010	1	0010100100	4
1010101000	5	0010100010	19
1010100100	1	0010100001	2
1010100010	5	0010100000	121
1010100000	27	0010010000	51
1010010000	1	0010001010	9
1010001010	3	0010001001	3
1010001000	23	0010001000	81
1010000100	5	0010000100	14
1010000010	20	0010000010	52
1010000001	3	0010000001	16
1010000000	358	0010000000	1781
1001000000	58	0000110000	27
1000110000	3	0000101100	2
1000101000	7	0000101010	10
1000100010	2	0000101001	1
1000100000	37	0000101000	56
1000100000	1	0000100100	11
1000010000	18	0000100011	1
1000001100	2	0000100010	31
1000001010	7	0000100001	5
1000001000	34	0000100000	419
1000000100	19	0000001100	34
1000000010	19	0000001011	2
1000000001	7	0000001010	83
1000000000	1572	0000001001	19
0011000000	167	0000001000	427
0010110000	3	0000000011	30
0010101100	1	0000000010	315

S_j, and Anderson and Sterling (1974) used the ratio $\hat{F}_j = \hat{\phi}_j / \hat{S}_j$ to estimate the probability that ducks banded in year j were still at risk of capture in year $j+1$.

Mardekian and McDonald (1981) proposed a method for analyzing joint band-recovery and live-recapture data that exploited the modeling approach and computer programs of Brownie et al. (1978, 1985).

Following release at a specific capture time, an animal may fall into one of three mutually exclusive categories:

1. the animal is found dead or killed and its band is reported (terminal recovery);
2. the animal returns to the banding site in a subsequent year and is captured; and
3. the animal is never seen again and its band is never reported.

Because animals can be recaptured several times, the full model allowing for all such recapture–recovery paths is difficult to write out. Mardekian and McDonald (1981) avoided this complication by ignoring intermediate recapture data and modeling the last encounter of each animal. The last encounter of each animal is either recovery of the band or the last live recapture. Mardekian and McDonald (1981) showed that $\hat{S}_i$ from the tag-return model fitted to the last encounter is an estimate of the survival probability between i and $i+1$.

The advantage of the Mardekian-McDonald analysis is the simplicity with which it can be carried out. The estimates can be obtained in a straightforward manner utilizing the procedures and software of Brownie et al. (1985). However, because intermediate live-recaptures are ignored, the Mardekian-McDonald survival rate estimator is not fully efficient. Another drawback of the Mardekian-McDonald analysis is that it is correct only if it is assumed that there is random or no emigration of animals. Under permanent or Markovian emigration it can be shown to lead to biased estimators (Barker 1995). When emigration is permanent, the estimator of Anderson and Sterling (1974) can be used instead, but it will also lead to biased estimates under Markovian emigration. Burnham (1993) established the foundation for an efficient analysis of the joint live-recapture and tag-return model. The basic model developed by Burnham (1993) combines the CJS model with the time-specific band-recovery model M1 of Brownie et al. (1985). That is, recapture, recovery, survival, and movement probabilities are time-specific and there is no heterogeneity. Burnham's (1993) model offered the alternative movement assumptions of random or permanent emigration.

As in the CJS model, the likelihood function for the joint model is proportional to the joint probability of the observed recapture and recovery events for the animal. It is known that the 76 animals with the encounter history 1100000000 in the paradise shelduck study died between January 1986 and January 1987, and given that they died, their bands were found and reported. Therefore, the joint probability Pr(1100000000 | first release in 1986) = $(1 - S_1)r_1$. Similarly, it is known that the 16 animals with the history 1011000000 survived between 1986 and 1987, remained within the local population that was at risk of capture in January

1987 and were caught, and then they died between 1987 and 1988 and were reported dead. The joint probability of these events is given by

$$\Pr(1011000000 \mid \text{first released in } 1986) = S_1 F_1 p_2 (1 - S_2) r_2$$

Under both random and permanent movement assumptions, joint live-recapture and dead-recovery data can be summarized using combined re-capture and recovery array known as an extended m-array (table 7.2). An advantage of this summary using sufficient statistics is that it provides a basis for efficient goodness-of-fit tests for these models. The extended m-array is a simple generalization of a band-recovery array for model M_1 of Brownie et al. (1985). From the encounter histories, the extended m-array can be constructed by considering the first reencounter of an animal after release following a live recapture. In a study with k capture occasions, and where recoveries are obtained up until occasion g $(g \geq k)$, each of the R_j marked animals released at capture occasion j can next be encountered either by a live recapture in one of the samples at time $j+1, \ldots, k$ or by dead recovery in one of the intervals between the capture occasions at times $j, j+1, \ldots, g$. If the number of animals released at time j and next encountered by live recapture at time h is represented by $m_{jh}^{(l)}$, and the number of animals released at time j and next encountered by dead recovery at time h by $m_{jh}^{(d)}$, then the data can be represented as in table 7.2. In this summary, the animals with the encounter history 1100000000 contribute to just $m_{11}^{(d)}$. The animals with the history 1011000000 first contribute to $m_{12}^{(l)}$ through the live recapture at time 2.

TABLE 7.2

Extended m-array illustrating the combined recapture and recovery data necessary for a joint analysis of live-recapture and tag-recovery data

Number released	Next encountered by recapture				Next encountered by dead recovery						Total
	$j=2$	3	$\cdots$	k	$j=1$	2	$\cdots$	$k-1$	k	$\cdots$	g	
N_1	$m_{12}^{(l)}$	$m_{13}^{(l)}$	$\cdots$	$m_{1k}^{(l)}$	$m_{11}^{(d)}$	$m_{12}^{(d)}$	$\cdots$	$m_{1k-1}^{(d)}$	$m_{1k}^{(d)}$	$\cdots$	$m_{1g}^{(d)}$	$m_{1.}^{(.)}$
N_2		$m_{23}^{(l)}$	$\cdots$	$m_{2k}^{(l)}$		$m_{22}^{(d)}$	$\cdots$	$m_{2k-1}^{(d)}$	$m_{2k}^{(d)}$	$\cdots$	$m_{2g}^{(d)}$	$m_{2.}^{(.)}$
$\vdots$			$\ddots$	$\vdots$			$\ddots$	$\vdots$	$\vdots$	$\cdots$	$\vdots$	$\vdots$
N_{k-1}				$m_{k-1k}^{(l)}$				$m_{k-1k-1}^{(d)}$	$m_{k-1k}^{(d)}$	$\cdots$	$m_{k-1g}^{(d)}$	$m_{k-1.}^{(.)}$
N_k									$m_{kk}^{(d)}$	$\cdots$	$m_{kg}^{(d)}$	$m_{k.}^{(.)}$
Total	$m_{.2}^{(l)}$	$m_{.3}^{(l)}$	$\cdots$	$m_{.k}^{(l)}$	$m_{.1}^{(d)}$	$m_{.2}^{(d)}$	$\cdots$	$m_{.k-1}^{(d)}$	$m_{.k}^{(d)}$	$\cdots$	$m_{.g}^{(d)}$	

Note. The extended m-array reveals the sufficient statistics to perform the analysis of combined data and provides a basis for efficient goodness-of-fit tests for these models.

TABLE 7.3

Extended m-array illustrating the joint live-recapture and tag-recovery data from the paradise shelduck (*Tadorna variegata*) banding study in the Wanganui Region, New Zealand, 1986–1990

Year (j)	R_j	Next encountered by live recapture				Next encountered by dead recovery					Total
		1987	1988	1989	1990	1986	1987	1988	1989	1990	
1986	2330	468	50	43	19	76	58	18	19	7	758
1987	2825	0	222	119	72	0	183	52	19	19	686
1988	835	0	0	116	58	0	0	33	16	7	230
1989	843	0	0	0	119	0	0	0	39	23	181
1990	613	0	0	0	0	0	0	0	0	33	33
Total		468	272	278	268	76	241	103	93	89	

Note. The sum of the values in bold represents z_{1987}, where z_j is the number of marked animals in the population known to have been alive at occasion j but not captured or killed until after j.

Because they were released at time 2 they are included in the R_2 animals released at time 2, and then they contribute to $m_{22}^{(d)}$. The extended m-array for the Wanganui paradise shelduck study is given in table 7.3.

Under the assumption that the fates of each animal in a release cohort are independent, the data in the extended m-array can be modeled using multinomial distributions. Each release cohort represents an independent multinomial experiment. Under random emigration, the probabilities associated with the statistics $m_{ij}^{(l)}$ in Table 7.2 are given by

$$
\pi_{jh}^{(l)} = \frac{E[m_{jh}^{(l)}|R_j]}{R_j} = \begin{cases} S_j F_j p_{j+1} & h = j+1 \\ S_{h-1} F_{h-1} p_h \prod_{t=j}^{h-2} S_t (1 - F_t p_{t+1}) & h = j+2, \ldots, k \end{cases}
$$

(7.1)

and

$$
\pi_{jh}^{(d)} = \frac{E[m_{jh}^{(d)}|R_j]}{R_j} = \begin{cases} (1 - S_j) r_j & h = j \\ (1 - S_h) r_h \prod_{t=j}^{h-1} S_t (1 - F_t p_{t+1}) & h = j+1, \ldots, g \end{cases}
$$

(7.2)

To see that this provides the correct probability structure under random emigration, consider an animal released at occasion j and next captured at time $j + 2$. It is known that it survived from occasions j to $j + 1$ and from occasions $j+1$ to $j+2$, and that it was caught at occasion $j+2$. What is unknown is whether it was at risk of capture at time $j+1$. However, there are just two mutually exclusive possibilities. The first possibility is that the animal was at risk of capture at time $j+1$, but not caught, which occurs with probability $F_j(1-p_{j+1})$. The animal then remained at risk of capture and was caught at time $j+2$, which has probability $F_{j+1}p_{j+2}$. The second possibility is that the animal temporarily emigrated between time j and $j+1$, which occurs with probability $(1-F_j)$, and then it returned to the local study population and was caught, which has probability $F_{j+1}p_{j+2}$. The probability that it was next captured at $j+2$ given that it was released at j is therefore

$$S_j S_{j+1}[F_j(1-p_{j+1}) + (1-F_j)]F_{j+1}\,p_{j+2}$$

which reduces to

$$S_j[(1-F_j p_{j+1})S_{j+1}F_{j+1}\,p_{j+2}$$

Under permanent emigration the situation is complicated by the fact that although animals can permanently emigrate from the local study population and no longer be at risk of capture, they are exposed to dead recovery throughout their range. Under permanent emigration the multinomial cell probabilities are given by

$$\pi_{jh}^{(l)} = \begin{cases} S_j F_j p_{j+1} & h = j+1 \\ S_{h-1}F_{h-1}p_h \displaystyle\prod_{t=j}^{h-2} S_t F_t (1-p_{t+1}) & h = j+2, \ldots, k \end{cases} \tag{7.3}$$

and

$$\pi_{jh}^{(d)} = \begin{cases} (1-S_j)r_j & h = j \\ (1-S_h)r_h \gamma_{jh} S_j \cdots S_{h-1} & h = j+1, \ldots, g \end{cases} \tag{7.4}$$

where

$$\gamma_{jh} = \begin{cases} 1 & j = h \\ 1 - F_j[1-(1-p_{j+1})]\gamma_{j+1,h} & j = 1, \ldots, k; \quad h = j+1, \ldots, k \\ \gamma_{jk} & j = 1, \ldots, k; \quad h = k+1, \ldots, g \end{cases}$$

$$\tag{7.5}$$

and is the probability that an animal alive and at risk of capture at capture occasion j, and alive at h, is not captured between occasions j and h. Computation of the γ_{jh} parameters is done recursively for any h by iterating backward from $j = h, h - 1, \ldots, 1$ (Burnham 1993).

Because the multinomial cell probabilities differ between the two forms of emigration in the joint model they do not have the same likelihood and the two types of emigration, random or permanent, can be distinguished. However, under random emigration the parameters F_j and p_{j+1} ($j = 1, \ldots, k - 1$) are confounded as they always appear in the model as the product $F_j p_{j+1}$. Under permanent emigration, the joint model allows estimation of $F_1, \ldots, F_{k-2}$ but the pair F_{k-1} and p_k are confounded.

Parameter Estimation

Under permanent emigration, closed-form maximum likelihood estimators (MLEs) and their variances do not exist for Burnham's (1993) joint model except in very restrictive cases. Instead, numerical solutions to the likelihood function must be found using a suitable software package. Under random emigration closed-form MLEs do exist and in terms of the notation in table 1.1 are given by

$$\hat{S}_j = \frac{\hat{M}_{j+1}}{\hat{M}_j - m_{\cdot j}^{(l)} + R_j}$$

$$\hat{p}_j = \frac{m_{\cdot j}^{(l)}}{\hat{M}_j}$$

and

$$\hat{r}_j = \frac{\hat{f}_j}{1 - \hat{S}_j}$$

where

$$\hat{M}_j = \frac{R_j z_j}{m_{j\cdot}^{(\cdot)}} + m_{\cdot j}^{(l)}$$

$$\hat{f}_j = \frac{m_{\cdot j}^{(d)}}{\hat{M}_j - m_{\cdot j}^{(l)} + R_j}$$

and z_j is the number of marked animals in the population at occasion j that were not caught at j, so that $z_1 = 0$ and $z_j = z_{j-1} + m^{(\cdot)}_{j-1\cdot} - m^{(l)}_{\cdot j}$. The dot in place of an index indicates summation across the replaced index. For example, $m^{(\cdot)}_{j\cdot} = m^{(l)}_{jj+1} + \cdots + m^{(l)}_{jk} + m^{(d)}_{jj} + \cdots + m^{(d)}_{jt}$, and is the number of the R_j animals in the jth release cohort that were ever encountered again. Similarly, $m^{(l)}_{\cdot j}$ is the total number of live recaptures in the jth interval and $m^{(d)}_{\cdot j}$ is the total number of dead recoveries between j and $j+1$. These summary statistics are all easily obtained from the modified m-array (table 7.3). The z_j also can be obtained from the m-array as the sum of the values in the upper submatrices (for capture and dead recovery) above the occasion in question. Hence, the $z_{1987} = 214$ is derived from the numbers shown in bold in table 7.3.

These summary statistics are presented more compactly in table 7.4, where we find $\hat{M}_{1986} = 0$ (the number of marked animals immediately before the first sampling occasion is always zero) and $\hat{M}_{1987} = (2825 \times 214)/686 + 468 = 1349.268$, which gives

$$\hat{S}_{1986} = \frac{1349.268}{0 - 0 + 2330} = 0.579, \qquad \hat{p}_{1987} = \frac{468}{1349.268} = 0.347,$$

$$\hat{f}_{1986} = \frac{76}{0 - 0 + 2330} = 0.0326, \quad \text{and} \quad \hat{r}_{1986} = \frac{0.0326}{1 - 0.579} = 0.077$$

A full set of estimates computed from the data in table 7.4 is given in table 7.5.

Estimators for the sampling variances are

$$var(\hat{S}_j) = S_j^2 \left[\frac{1}{m^{(\cdot)}_{\cdot j}} - \frac{1}{R_j} + (1 - p_{j+1})^2 \left(\frac{1}{m^{(\cdot)}_{\cdot j+1}} - \frac{1}{R_{j+1}} \right) + p_{j+1}^2 \left(\frac{R_{j+1} - m^{(\cdot)}_{\cdot j+1}}{m^{(\cdot)}_{\cdot j+1}} \right) \right.$$

$$\left. \times \left(\frac{1}{m^{(l)}_{\cdot j}} - \frac{1}{m^{(l)}_{\cdot j} + z_j} \right) + \frac{1}{m^{(\cdot)}_{\cdot j} + z_l - m^{(d)}_{\cdot j}} - \frac{1}{m^{(\cdot)}_{\cdot j} + z_i} \right]$$

$$var(\hat{p}_j) = p_j^2 (1 - p_j)^2 \left(\frac{1}{m^{(\cdot)}_{\cdot j}} - \frac{1}{R_j} + \frac{1}{m^{(l)}_{\cdot j+1}} + \frac{1}{z_j} \right)$$

and

$$var(\hat{r}_j) = r_j^2 \left[\frac{var(\hat{f}_j)}{f_j^2} + \frac{var(\hat{S}_j)}{(1 - S_j)^2} + \frac{2cov(\hat{f}_j, \hat{S}_j)}{f_j(1 - S_j)} \right]$$

TABLE 7.4

Summary statistics for the joint live-recapture and tag-recovery data for paradise shelduck (*Tadorna variegata*) banded in the Wanganui Region, New
Zealand, 1986–1990, where R_j is the number of marked animals released at
occasion j.

Sample (j)	R_j	$m_{j.}^{(.)}$	$m_{.j}^{(l)}$	$m_{.j}^{(d)}$	z_j
1986	2330	758		76	0
1987	2825	686	468	241	214
1988	835	230	272	103	387
1989	843	181	278	93	236
1990	613	33	268	89	56

Note. The statistic z_j (the number of marked animals in the population at j that were not
caught at j but that were caught later) is found by $z_j = z_{j-1} + m_{j-1,.}^{(.)} - m_{.,j-1}^{(d)} - m_{.j}^{(l)}$.

where

$$var(\hat{f}_j) = f_j^2 \left(\frac{1}{m_{.j}^{(.)}} - \frac{1}{R_j} + \frac{1}{m_{.j}^{(d)}} - \frac{1}{m_{.j}^{(.)} + z_j} \right) \qquad (7.13)$$

It is useful to note that $Var(\hat{S}_j)$ is the same as under the CJS model but
with the statistics R_j, $m_{.j}^{(.)}$, and z_j augmented by the dead recoveries, and
with the addition of the term $1/(m_{.j}^{(.)} + z_j - m_{.j}^{(d)}) - 1/(m_{.j}^{(.)} + z_j)$. Because of this
extra term, simply plugging the augmented statistics into the CJS vari-
ance estimator will underestimate the true sampling variance. The extra
term can be thought of as a penalty term that adjusts the variance estimator to account for the estimation of the reporting parameter r_j.

TABLE 7.5

Parameter estimates for the joint live-recapture and tag-recovery data for paradise shelduck (*Tadorna variegata*) banded in the Wanganui Region, New
Zealand, 1986–1990 under random emigration

	$\hat{S}_j$	$SE(\hat{S}_j)$	$\hat{p}_j$	$SE(\hat{p}_j)$	$\hat{r}_j$	$SE(\hat{r}_j)$
1986	0.579	0.029	—	—	0.077	0.010
1987	0.453	0.027	0.347	0.020	0.119	0.010
1988	0.615	0.051	0.162	0.013	0.119	0.022
1989	0.674	0.118	0.202	0.018	0.147	0.058
1990			0.205	0.037	—	—

Assumptions

Key assumptions of the above model are that fates of animals are independent and that parameters differ only by time and not among individual animals or according to the capture history. In addition, it is important that the recapture process operates over a short period of time during which the population can be considered closed. Note that the population does not need to be closed during the dead-recovery interval, which may be the entire time between live-capture occasions. If specific effects are believed present that mean the assumptions do not hold, the model may be modified accordingly. There have been a number of developments extending the above models to account for specific structure in the analysis. Some of these are introduced in the next section. It is also desirable to have a method for assessing how well a particular model accounts for variation in the data.

Goodness-of-fit tests comprising two sets of contingency tables similar to TEST2 and TEST3 of Burnham et al. (1987) can be constructed for these models. The theoretical basis of these tests is given by Barker (1997) for the joint models and extends the work of Pollock et al. (1985) and Brownie and Robson (1983) for live-recapture and tag-resighting studies and by Brownie et al. (1985) for tag-recovery models.

The first set tests whether the next encounter following the current live capture depends on the previous capture history and generalizes TEST3 of program RELEASE (Burnham et al. 1987). The test is valid under random, permanent, or Markov emigration. For the 1988 release cohort in the paradise shelduck study, the full contingency table for the first test component is given in table 7.6. This table is found by classifying each of the animals that were released in 1988 first according to their capture history before the 1988 release (i.e., identification of subcohorts), and second according to when and how they were next encountered. Although the recapture and recovery patterns for each subcohort appear similar in table 7.6, more of the newly marked birds were never seen again (433/577 = 0.75) than the birds that had been caught at least once before

$$\frac{126 + 37 + 27}{188 + 50 + 39} = \frac{190}{277} = 0.69$$

(chi-square statistic = 8.17 with 4 df, $p = 0.004$). There are several reasons why this could happen, including temporary trap shyness resulting from first capture, an age effect (since newly marked cohorts tend to include younger birds), or the presence of transients (Pradel et al. 1997a) in the population.

TABLE 7.6

Contingency table for the 1988 release cohort testing whether the classification of birds according to when they were next encountered depends on their capture history prior to release in 1988

Encounter history before release	Number released	Next encounter					Never seen again
		Recapture		Tag recovery			
		4	5	3	4	5	
00001	577	69	32	27	11	5	433
00101	188	34	19	3	4	2	126
10001	50	8	2	3	0	0	37
10101	39	6	5	0	1	0	27

Under random emigration, a second set of goodness-of-fit tests can be constructed that generalize TEST2 of RELEASE (Burnham et al. 1987). These are found by partitioning the $m_{\cdot j}^{(\cdot)} + z_j$ animals known to be alive immediately following capture occasion j first according to whether or not the animals were captured at time $j + 1$ and second according to how and when the animals were next encountered after time $j + 1$.

The full contingency table for this test component constructed from marked paradise shelduck known to be alive immediately after the 1988 sample is given in table 7.7. The counts in the contingency table suggest that birds not caught in 1988 have recapture and recovery patterns similar to those of birds caught in 1988, although there is weak evidence against the model (chi-squared statistic = 8.73 with 4 df, $p = 0.07$). Possible reasons for this test indicating lack of fit include a temporary effect of capture on the pattern of recapture or dead recovery and failure of the random emigration assumption.

An overall test statistic is obtained by pooling results from all the contingency tables constructed as above. For the paradise shelduck study there are seven contingency tables in all, and summing across these yields a chi-squared statistic of 56.95 with 16 df ($p < 0.0001$), providing strong evidence that the model is not adequate.

If emigration is permanent or Markovian, the second set of goodness of fit tests generalizing TEST2 described above are not valid. Instead, a test can be constructed by comparing the observed and expected values of the m-array. Each row of the m-array corresponds to a particular release cohort and yields a test-statistic. An overall test statistic is obtained by summing across each row of the m-array. Although the same procedure can be followed for random emigration, the contingency table test described above will be more efficient.

TABLE 7.7

Contingency table for the $m_{\cdot 3}^{(\cdot)} + z_3$ marked birds known to be alive immediately after the 1988 sampling occasion testing whether recapture or recovery after the 1988 sample depends on whether or not the bird was caught in the 1988 sample

| | Next encounter | | | | | |
| | Recapture | | Dead recovery | | | |
	89	90	88	89	90	Total
Last caught before 1988 (Z_3)	162	91	70	38	26	387
Caught in 1988 ($m_{\cdot 3}^{(\cdot)}$)	116	58	33	16	7	230
Total	278	149	103	54	33	617

7.4 More General Models

If key model assumptions are violated then biased estimation will likely result. In addition, sampling variance estimates may be biased. Structural inadequacy of the model may be apparent from goodness-of-fit testing, or there may be theoretical reasons for believing that the model is inadequate. The joint model of Burnham (1993) has been generalized in two key ways.

Catchpole et al. (1998) relaxed the assumptions to allow both age- and time-dependent parameters. A more general extension of the model is included in the computer program MARK (White and Burnham 1999; White et al. 2001). In addition to age and time dependence, MARK allows the parameters in the model to depend on individual covariates and also allows environmental covariates that can be used to explain time variation in the parameters.

The joint live-recapture, live-resighting, and tag-recovery model of Barker (1997) and Barker et al. (2004) covers the general situation where there are live recaptures, live resightings from any time between the capture occasions, and dead recoveries that occur any time between marking occasions. The joint live-recapture, live-resighting, and tag-recovery model provides three important generalizations of tag-resighting and tag-return models. First, it allows information from recaptures of marked animals during the tagging occasions to be included in the analysis. Second, it relaxes the assumption that the population is closed during the resighting period. Third, it allows Markovian temporary emigration as well as the special cases of permanent and random emigration.

In addition to the parameters of Burnham's joint model, Barker's model introduces two more resighting parameters. The first is ρ_j, which is the probability a marked animal alive at sampling occasions j and $j + 1$ is resighted between j and $j + 1$. The second is ρ_j', which is the probability that an animal that dies between occasions j and $j+1$, but that was not found and reported, was resighted alive between j and $j+1$ before it died.

The CJS model, the tag-recovery model, and Burnham's joint live-recapture dead-recovery model are special cases that arise by making appropriate constraints on the parameters. For example, constraining the live resighting parameters ρ_i and ρ_i' to equal zero leads to Burnham's (1993) joint model but with Markovian emigration. Another special case arises by assuming no live recaptures but resighting over an open period between the capture occasions in which newly marked animals are released. The survival parameters in this model can be estimated and it represents a model that can be validly applied in a tag-resighting study with an open resighting interval.

7.5 Model Fitting and Assessment

As for Burnham's joint model, closed-form maximum likelihood estimators exist only under the random emigration assumption and are given by Barker (1997). Under permanent or Markov emigration numerical methods must be used to find solutions to the likelihood equations. The model can be fitted using the likelihood constructed directly from the encounter histories (Burnham 1993).

The likelihood function is constructed in a similar manner to that for Burnham's (1993) joint model. Consider a bird in the paradise shelduck study that was released in 1986, resighted alive between 1986 and 1987, that was captured in 1987, then reported shot during the 1987 hunting season. This bird would have the encounter history 1211000000 with the associated probability

$$\Pr(1211000000 \mid \text{first released in 1986}) = S_1\rho_1 F_1 p_2(1 - S_1)r_1$$

Because we know that this animal was alive in 1987, the resighting event is coded with the parameter ρ_1, the probability an animal alive at times 1 and 2 is resighted alive in this interval. The parameter ρ_j', the probability that an animal that dies between times j and $j+1$ without being reported dead is resighted before it dies, enters into the likelihood function only for animals that are never seen again following a live resighting. For example, consider a bird released in 1990, the last capture occasion, that was resighted alive between 1990 and 1991. It is not known whether this bird was alive in January 1991; therefore, the probability of the encounter history is given by

$$\Pr(0000000012 \mid \text{first released in 1990}) = S_5\rho_5 + (1 - S_5)(1 - r_5)\rho_5'$$

Direct modeling of the encounter histories is very flexible, because parameters can be modeled using explicit functions of covariates that may vary over time, or by individuals, using a generalized linear model formulation. If θ_{ij} is a parameter unique to individual i at time j, then covariates can be included in the model by using the expression $\eta_{ij} = \beta_0 + \beta_1 \mathbf{x}_j + \beta_2 \mathbf{x}_i$ where $\eta_{ij} = h(\theta_{ij})$ is the link function (McCullagh and Nelder 1989), β_1 is a vector of effects for time-specific covariates $\mathbf{x}_j$, and β_2 is a vector of effects for individual-specific covariates $\mathbf{x}_j$. The generalized linear model formulation offers much flexibility. For example, with a little thought joint models that allow age-dependence and short-term marking effect can be fitted.

To introduce a short-term marking effect on survival using individual covariates, survival is modeled as $h(S_{ij}) = \beta_{0j} + \beta_{1j} x_{ij}$, where $x_{ij} = 1$ if animal i is marked and released for the first time at time j, or 0 otherwise. This approach allows the models of Pollock (1975) and Brownie and Robson (1983) to be extended to incorporate live recaptures and tag recoveries, and the models of Brownie et al. (1985) to be extended to include live recaptures. Barker (1999) developed explicit estimators and tests for age dependence and short-term marking effect but only under the random emigration assumption.

A very large number of models can be fitted to a data set by including those models that are constructed by placing restrictions on parameters. In a study where animals are classified according to sex, parameters can be constant, sex-specific, time-specific, or sex- and time-specific. With seven parameter types in the model for the joint analysis of live-recapture, live-resighting, and dead-recovery data, 16,384 models can be constructed by considering all combinations of this nature. Because the resighting parameters ρ_j and ρ_j' model similar events, as do the movement parameters F_j and F_j', it may be reasonable to consider just models where the same effects apply to each member of the pair. Even so, there are still 1024 models that can be constructed in this case.

The very large number of potential models compared to the CJS or tag-return models means that more realistic models can be constructed, but it also emphasizes the need for sensible model-selection strategies and efficient model-selection algorithms, such as those based on Akaike's Information Criterion (chapter 1). Because it is not practical to fit all possible models it is important to think about the effects that are likely to be present before fitting models. For example, in most studies researchers have little or no control over the tag-recovery and resighting processes. In this case time-varying reporting and resighting parameters will probably be needed in the model. In species that are not sexually dimorphic and where males and females behave similarly there may not be a sex effect on reporting or resighting parameters.

A related issue is that joint models have many more parameters than their live-recapture or tag-recovery counterpart. In the paradise shelduck example described below where there are just five capture occasions and no separation of birds by sex, the fully time-dependent model with no individual covariates has 31 parameters. The equivalent CJS model with six sampling periods has just eight parameters. It might be thought that the need to estimate extra parameters will reduce the precision of the survival rate estimates. However, as shown by Barker and Kavalieris (2001), this is not the case. The CJS and tag-recovery models are special cases of the joint models, and the survival probability can be estimated under both models without the extra information provided by the joint model. The extra parameters are required only to model the additional data. If there is little useful information in the additional data, the extra parameters will be poorly estimated, but not the survival rates.

Goodness-of-Fit

The tests described above for the analysis of live-recapture and dead-recovery data were developed by Barker (1997) for the more general situation where there are live resightings as well. Accordingly, the two components described above differ slightly in the more general case. Importantly, the m-array represents a set of sufficient statistics only under random emigration for this more general model so the tests described here are valid only in the case of random emigration. As in the tests above, the first set of tests is used to assess whether the next encounter following the current live encounter depends on the encounter history to date. Including live resightings means there are now a greater number of histories possible before and after the current encounter.

The second set of tests partitions the $m^{(\cdot)}_{\cdot j} + z_j$ animals known to be alive immediately following capture occasion j according to whether they were (1) caught at j, (2) resighted alive between occasions $j - 1$ and j, but not caught at j, and (3) not resighted between $j - 1$ and j, and not caught at j. Because Burnham's (1993) model does not include live resightings, this partitioning reduces to whether or not the animals were recaptured at time $j + 1$ in his model. The second partition is according to how and when the animals were next encountered after time $j + 1$. This test examines whether an encounter between occasions j and $j + 1$ has longer-term effects on the probability of later encounters, and generalizes TEST2 of RELEASE (Burnham et al. 1987).

In addition to the two components described above a third set of contingency tables are based on two partitions of the $m^{(l)}_{\cdot j} + z_j$ marked animals that are known to be alive at sample time $j - 1$ and just before sample j. The first partition is according to whether or not they were captured at j. The second is according to whether they were resighted alive between occasions $j - 1$ and j. The cross-classification arising from the two partitions generates contingency tables that test whether resighting has a short-term effect on the probability of recapture. This test is not relevant for Burnham's (1993) model because the model does not involve live resightings.

The usefulness of these tests is questionable because they test the fit of the joint models only under random emigration and not in the more general case of Markov emigration. However, the comparison of expected and observed values in the contingency tables may still be useful for highlighting model inadequacies. If emigration is not random, a goodness-of-fit test can be constructed using a parametric bootstrap procedure provided there are no individual covariates. In the parametric bootstrapping procedure data are simulated under the assumed model using the parameter estimates for the parameter values. The simulation process is repeated a large number of times, and each time the value of the deviance statistic (equation 1.4 with L_2 computed for a saturated model) is recorded. If few of the simulated deviance values exceed the value for the observed data, this is taken as evidence that the data are inconsistent with the assumed model.

A philosophical objection to a goodness-of-fit test is that there is never complete faith in assumptions. The model is just an approximation to reality, and will fail the test given enough data. An alternative approach is to have a measure of the extent to which the model fails to represent the data. The overall goodness-of-fit statistic for the whole model, however it is calculated, divided by its degrees of freedom is one such estimate of the extent of lack of fit. This is the statistic $\hat{c}$ of equation 1.8.

Two important types of failure of the model are when, first, the fate of one animal influences the fate of another, and, second, when parameters are not the same for each animal. If fates are not independent, or if parameters vary randomly from animal to animal, the data will often be overdispersed relative to the model. If the data are overdispersed, the parameter estimators are unbiased but model-based standard errors will be too small. Quasi-likelihood (section 1.4) has been suggested as a method for approximating the true sampling model for the data, where the variances of the parameters are the same as for the model with no overdispersion except that they are multiplied by the estimated scaling constant $\hat{c} > 1$.

With mark–recapture models $\hat{c}$ can be found in one of two ways. The first is to use an explicit goodness-of-fit test, as outlined above, to find an overall χ^2 statistic and its associated degrees of freedom, d. The estimator $\hat{c} = \chi^2/d$ then provides an estimate of c. The second approach is to use the parametric bootstrap method. The approach recommended by White et al. (2001) is to divide the observed deviance for the data by the mean of the simulated deviances. The mean of the simulated deviances represents the expected value of the deviance under the null hypothesis that the model is correct. Therefore, the observed deviance divided by the expected deviance is an estimate of the overdispersion in the data.

With overdispersion, the variances of the parameter estimates are underestimated by a factor of $1/c$. Therefore, the estimate $\hat{c}$ can be used to scale up the sampling variances to correct for overdispersion.

7.6 Tag Misreads and Tag Loss

Because the joint models incorporate information provided by members of the public, data quality is an important issue. Misreading of tag numbers may be more common than in studies in which a limited number of project personnel perform the work. Similarly, misreading of tags is likely to be more of a problem in resighting studies than in studies where animals are captured, because errors in reading tags are more likely when tags are read at a distance. The effect of misread tags in a CJS-type study has been investigated by Schwarz and Stobo (1999). They showed that estimates of survival rate can be biased early on in the study, but that this bias tends to ameliorate later in the study. Schwarz and Stobo (1999) also developed a model for tag misreading that allows misread rates to be estimated using CJS data. Effects of tag loss can be similar to those of tag misreads. Large, high-visibility tags used in a resighting study may be prone to tag loss. Double tagging may help assess rates of tag loss (Fabrizio et al. 1999), but loss of tags from double-tagged animals may not be independent. This confounds efforts to assess loss rates (Bradshaw et al. 2000; Diefenbach and Alt 1998). The Schwarz and Stobo (1999) misread model also may be extended to studies involving double tagging. Pollock (1981), Chao (1988), Nichols et al. (1992a), and Fabrizio et al. (1999) have suggested a variety of methods to assess tag loss and how to mitigate its effects on estimation of population parameters.

7.7 Computing Considerations

Currently, the only widely available computer packages that allow the user to fit joint models efficiently are the packages SURVIV (White 1983) and MARK (White and Burnham 1999). SURVIV is useful in those cases where the model can be fitted to the extended m-array. Unfortunately, for the models discussed in this chapter, this restricts the use of SURVIV to the joint live-recapture, dead-recovery models based on Burnham's (1993) model, or the joint live-recapture, live-resighting, and dead-recovery model based on Barker's (1997) model under the assumption of random emigration.

One advantage of SURVIV is that it automatically includes a goodness-of-fit test. This test, however, is not fully efficient for the models considered in this chapter because it does not include the first component of the tests described above, that component can be thought of as assessing the sufficiency of the m-array. Disadvantages of SURVIV are, first, that the user needs to code the algebraic structure of the model by hand before the model can be fitted. Equations 7.1 to 7.5 in Burnham (1993) can help with this. Second, SURVIV is not suitable if the user wishes to include individual covariates in the analysis.

Program MARK (White and Burnham 1999) fits the joint models described above using a likelihood function constructed directly from the encounter histories. A generalized linear model formulation is used to introduce individual covariates into the model. This process is very flexible as parameters can be indexed by release cohort, time, or individual. Because not all parameters can be estimated in a completely cohort- and time-specific model, constraints are introduced through a parameter index matrix (PIM). Model assessment in MARK for joint live-recapture, live-resighting, and dead-recovery models is carried out using parametric bootstrapping.

By fixing parameters it is possible to fit restricted versions of the joint model. For example, by constraining the resighting parameters ρ_j and ρ_j' to equal zero, the joint model for live recaptures, live resightings, and tag recoveries can be used to fit Burnham's joint model. This is currently the only way to fit a Markovian emigration model in MARK to joint live-recapture and tag-recovery data. If the capture probabilities p_j and the tag-reporting probability r_j are constrained to equal zero, it is possible to fit a tag-resighting model that allows the population to be open during the resighting period.

The permanent emigration model can be fitted using MARK by setting all the F_j' equal to zero. Under random emigration, $F_j' = F_j$, but F_j is now confounded with the capture probability p_{j+1}. A useful computational trick is to set the constraint $F_j = 1$. Because F_j and p_{j+1} always appear in the product $F_j p_{j+1}$, this ensures that the numerical procedure will find estimates of p_{j+1} that correspond to $F_j p_{j+1}$. Under the constraint $F_j = 1$, the parameters F_j will make no contribution to the likelihood and so can be arbitrarily constrained to any real value.

In many studies data will be too sparse to allow all parameters to be estimated. MARK will successfully maximize the likelihood in this situation, but the estimates of the inestimable parameters will not be unique, which is usually indicated by a very large estimated standard error. MARK attempts to determine which parameters are uniquely estimated using numerical methods, and usually does so correctly if the default sin link is used. For other link functions, the parameter count may not be correct and this usually happens when some of the parameter estimates are at their boundary value, for example, a survival rate estimated to be 1.0.

TABLE 7.8
Model fitting summary from fitting the fully time-dependent Markov, permanent, and random emigration models to the paradise shelduck data

Model	Emigration	ΔAIC_c	Akaike weight	Number of parameters	Deviance
$S_t p_t r_t F_t F'_t = 0$	Permanent	0	0.608	16	123.50
$S_t p_t r_t F_t F'_t$	Markov	0.98	0.372	18	120.46
$S_t p_t r_t F_t = F'_t = 1$	Random	6.87	0.020	13	136.39

TABLE 7.9
Parameter estimates for the paradise shelduck data under random emigration from program MARK

	$\hat{S}_j$	SE	$\hat{p}_j$	SE	$\hat{r}_j$	SE
1986	0.579	0.028			0.077	0.01
1987	0.452	0.029	0.347	0.020	0.119	0.01
1988	0.615	0.052	0.162	0.013	0.119	0.022
1989	0.674	0.120	0.202	0.018	0.147	0.058
1990	0.479	23.754	0.205	0.037	0.103	4.716

Different models can be compared in MARK using the small-sample version of Akaike's Information Criterion (AIC_c) mentioned in chapter 1 for model selection. As an illustration, table 7.8 gives model fitting information for the three different movement models fitted to the paradise shelduck data. AIC_c indicates a preference for the permanent emigration model (Akaike weight = 0.608) or possibly the Markov emigration model (Akaike weight = 0.372) with much less support for the random emigration model. An estimate of the overdispersion scaling parameter c for the Markov emigration model, the most general of the three models, indicates considerable overdispersion ($\hat{c} = 1.83$). Also, of 1000 simulated deviance statistics, all were less than the deviance statistic for the data of 120.458. This suggests that there may be some structural inadequacies in the model for the paradise shelduck data. For comparison with the explicit solutions, the parameter estimates given by MARK for the random emigration model are given in table 7.9.

7.8 Chapter Summary

- An Example involving the banding of paradise shelducks is used to introduce situations where some marked animals are recovered alive and others are recovered dead. It is noted that it will be more efficient to produce parameter estimates from both types of recovery, rather than considering the information from either live or dead recoveries only.

- The effect of animal movement on the standard models for live and dead recoveries is discussed.
- The structure of the data from a study involving both live and dead recoveries is described.
- Simple early models that use both live- and dead-recovery data are discussed, including the Burnham model, which makes use of all of the available data and assumes that recapture, recovery, survival, and movement probabilities vary with time. The assumptions needed for estimation and goodness-of-fit tests are reviewed.
- A generalization of the Burnham model that allows parameters to vary both with age and time is discussed, as is another generalization that allows for temporary migration. Model fitting and assessment methods are discussed for the second of these models.
- The effects of misreading of tags and tag loss are discussed.
- Methods for computing estimates using data on live and dead recoveries are described.

Figure 7.2. Martha Tomeo creatively avoids being attacked by a neck-collared Canada goose (*Branta canadensis*), as she checks a nest near Anchorage, Alaska, 2000. (Photo by Jerry Hupp)

Eight

Multistate Models

CARL J. SCHWARZ

8.1 Introduction

The original modeling framework for capture–recapture studies assumed homogeneous behavior among animals and was concerned with estimating parameters, such as survival or abundance, for a single uniform population. In many situations, this is unrealistic because animals within a population are not homogeneous with respect to survival and catchability, and heterogeneity in these parameters can lead to biases in estimates.

In their key paper, Lebreton et al. (1992) developed a modeling framework based on partitioning populations into homogeneous subpopulations (groups) based on fixed, unchanging attributes, such as sex. These models and the AIC model selection framework (chapter 1) start with separate parameters for each group in the population and move toward models where some parameters are common across groups, e.g., male and females having common survival rates but unequal catchabilities.

The Lebreton et al. (1992) framework assumes that membership in the groups is fixed and does not change over time, which, in a way, is similar to stratification of populations. However, in some cases, it is desirable to let group membership change over time. These changes can be divided into two categories. First, some changes are predicable. If the population is divided into age classes, the movement of animals among the age classes is predictable and regular. Second, some changes occur randomly, such as migration among geographical areas, and the movement of animals is not predictable. Movement was the prime motivation for consideration of multistate models, but these models can be generalized to any situation where animals change state in an unpredictable fashion (Lebreton et al. 1999; Lebreton and Pradel 2002).

This chapter is divided into two sections. In the first, multistate models that estimate survival, movement among strata, and capture probabilities, but not abundance, are examined. These are generalizations of the

This work was supported by a Research Grant from the Canadian Natural Science and Engineering Research Council (NSERC).

Figure 8.1. Close-up photo of a "Coded Wire Tag" for permanent marking of many species of fish. Tags can be cut to length upon injection into the organism. (Courtesy of Northwest Marine Technology, Inc.)

Cormack-Jolly-Seber (CJS) models described in chapters 3 and 5. Animals are marked, released into the population, and followed over time, but the process by which animals were initially marked is ignored. In the second part, models that also estimate abundance are introduced. In this part it is crucial that the process by which new animals are initially captured be properly modeled, as information from this process allows the estimation of abundance. This requires much more care in designing the study.

8.2 The Arnason-Schwarz Model

Introduction

Chapman and Junge (1956) and Darroch (1961) were among the first papers to take account of geographical sites in capture–recapture analyses of two-sample experiments. Arnason (1972, 1973) generalized these to three samples and Schwarz et al. (1993b) to the k-sample case. Arnason's pioneering work remained largely unused because of the lack of computer software, and because it seems to have limited robustness with its large number of parameters (Viallefont and Lebreton 1993). However, with the development of suitable software (MSSURVIV, Brownie et al. 1993; MARK, White and Burnham 1999; M-SURGE, Choquet et al. 2003), the application of these models became more widespread. Hestbeck et al. (1991) and Brownie et al. (1993) were among the first to consider multistate models in a movement context when examining the migration patterns of Canada Geese. Further work on estimating dispersal was reported in Nichols et al. (1993), Nichols and Kaiser (1999), and Schwarz and Seber (1999). Nichols et al. (1994) generalized stratification to non-geographic cases by using these models to examine changes among weight classes, and Lebreton and Pradel (2002) examine stratified models in a more general context.

Experimental Protocol and Data Collection

The experimental protocol for the Arnason-Schwarz (AS) multistate model is an extension of that used in the CJS model. Releases of tagged animals are made in states, which are labeled as 1, 2, 3, . . . , S. These are followed over time, and when an animal is recaptured, its state of recapture is recorded. After recording the tag information, the animal is released back to the population, although losses on capture are allowed. There are a total of k sampling occasions.

The raw data from such an experiment consists of a generalization of the history pattern used in CJS models. The history record is of length k, $h = (h_1, h_2, \ldots, h_k)$, where h_j is

0	if the animal was not recaptured or released at time j;
1, 2, 3, . . .	if the animal was initially released or was recaptured at time j in state 1, 2, . . . , S and was returned to the population; or
−1, −2, . . .	if the animal was recaptured at time j in state 1, 2, . . . and lost on captured (i.e., not returned to the population).

The first nonzero element of the history vector indicates the stratum of first release for each animal. For example, the history vector $h = (0, 2, 0, 1)$ would indicate an animal that was initially released in year 2 in stratum 2, not seen in the third year, and then seen in year 4 in stratum 1.

There are three sets of parameters in the AS multistate model. The first two sets are generalizations of the CJS capture and survival parameters to each stratum. The third set represents the movement of animals among strata.

As in the basic CJS model, subscripts on parameters refer to time. Superscripts will be used to represent strata. The recapture probabilities are denoted as p_j^s, the probability of being captured in stratum s at time j given that the animal is present in stratum s. The parameter ϕ_j^s represents the probability that an animal alive at time j in stratum s is alive at time $j + 1$ and in one of the (unspecified) strata 1, . . . , S. It is interpreted as an average survival probability for animals alive at time j in stratum s. As in the CJS model, survival refers to apparent survival, i.e., death and permanent emigration outside the study areas cannot be distinguished. Lastly, the movement parameters ψ_j^{st} represent the movement probabilities of an animal alive at time j in stratum s moving to stratum t at time $j+1$ conditional, upon surviving until time $j+1$. It is also assumed that

$$\sum_{t=1}^{S} \psi_j^{st} = 1$$

i.e., living animals must move to one of the strata in the study area, and animals that move off the study area are indistinguishable from those dying. Unless strong assumptions are made about behavior, it is not possible to model the timing of survival and movement. For example, one cannot distinguish between (1) movement taking place immediately after release with no real mortality during the movement, and all mortality taking place in the receiving stratum; or (2) animals remaining in the initial stratum for an extended period of time (and subject to mortality) followed by a movement just before the next sampling occasion. The movement probabilities are net; i.e., if an animal moves among many strata between time j and time $j+1$, only the initial and final stratum are of interest. Lastly, there is an implicit assumption that movements prior to time j do not affect the movement choices at time j (i.e., the assumption is of Markovian movement, as discussed in chapter 7).

In an experiment with k sample times and S strata, there are a total of $(k-1)S$ average survival parameters, $\phi_j^s, j = 1, \ldots, k-1$ and $s = 1, \ldots, S$; $(k-1)(S^2 - S)$ movement parameters, $\psi_j^{st}, j = 1, \ldots, k-1$ and $s = 1, \ldots, S$, but with $\psi_j^{s1} + \psi_j^{s2} + \cdots + \psi_j^{sS} = 1$; $(k-1)S$ capture parameters, $p_j^s, j = 2, \ldots, k$ and $s = 1, \ldots, S$; and finally $(k-1)S$ loss-on-capture parameter. This gives a total of $(k-1)(S^2 + 2S)$ parameters, compared to the $3(k-1)$ parameters in the CJS model.

The key step in any modeling framework is to express the probability of the observed history data in terms of the parameters of interest. Here, this step proceeds similarly to the illustration in chapter 1 for CJS models except that capture and survival parameters are stratum specific, and movement among strata must be accounted for. The probability expression for a particular history vector can be complex. It is fairly straightforward to write the contribution when an animal is sighted at two adjacent sample times. The complications arise when dealing with (1) a sighting gap (e.g., 2, 0, 1), (2) the fate of the animal after the last capture (e.g., ..., 1, 2, 0, 0, 0), and (3) losses on capture (e.g., ..., 1, 0, -2, 0, 0, 0). Generally, when an animal is unobserved at time j the stratum of membership is also unknown and so all possible movement paths between the observed locations must be modeled. For example, assuming that only two strata are present, the probability of the history $h = (1, 2, 0, 1)$ is

$$\phi_1^1 \psi_1^{12} p_2^2 [\phi_2^2 \psi_2^{21}(1 - p_3^1)\phi_3^1 \psi_3^{11} + \phi_2^2 \psi_2^{22}(1 - p_3^2)\phi_3^2 \psi_3^{21}] \, p_4^1$$

where $\phi_1^1 \psi_1^{12}$ is the probability that the animal released in stratum *1* at time *1* will survive to time 2 and be present in stratum *2*, p_2^2 is the probability that the animal alive at time *2* and in stratum *2* will be recaptured, and

$$\phi_2^2 \psi_2^{21}(1 - p_3^1) \, \phi_3^1 \psi_3^{11} + \phi_2^2 \psi_2^{22}(1 - p_3^2) \, \phi_3^2 \psi_3^{21}$$

is the overall probability that an animal alive in stratum *2* at time *2* will be alive at time *4* and in stratum *1*. The last term must account for the two possible paths from stratum *2* to stratum *1* between times 2 and 4, which are unobserved. Finally, p_4^1 is the probability that the animal alive at time *4* in stratum *1* was recaptured.

Once the probability for each history is defined, the overall probability of observing the data (the likelihood) can be determined using a multinomial distribution in much the same way as in simple models shown in chapter 1.

This model makes a number of assumptions in addition to those made for the CJS model:

    1. all data are recorded without error with respect to the stratum of recovery, resight, or recapture;

    2. there are equal movement probabilities for all animals in stratum *s* at time *j*;

    3. the movement probability and recapture probabilities do not depend on the past history of the animal; and

    4. recoveries are instantaneous.

Schwarz et al. (1993a) developed movement models for use when recoveries take place over a period of time such as in a fishery that are generalizations of Brownie et al. (1985).

Model Fitting and Selection

No simple formula exists for the maximum likelihood estimators, but Schwarz et al. (1993b, appendix A) present moment estimators. Standard numerical procedures can be used to maximize the likelihood function, to obtain estimates of the parameters, and to obtain estimated standard errors. Three software packages that can analyze the multistate model are MSSURVIV (Brownie et al. 1993), MARK (White and Burnham 1999), and M-SURGE (Choquet et al. 2003). Cooch and White (2001) give a detailed description on how to use MARK to fit multistrata models. Choquet et al. (2003) also has a manual demonstrating the use of M-SURGE.

It should be noted that there is a substantial risk of converging to local maxima of the likelihood (Lebreton and Pradel 2002). Choosing several different initial starting points and ensuring that these converge to the same values will serve as a check of the final model estimates. Similarly, any large change in the estimates of movement or catchability parameters between models with similar structure is an indication of lack of convergence.

As in the CJS model, not all parameters may be identifiable. In the most general model with capture, survival, and movement probabilities time dependent, the final survival, movement, and recapture probabilities are all confounded. Some parameters may also not be estimable because of sparse data, e.g., when there are no recoveries in any stratum at time i or no releases in a stratum at time i. Gimenez et al. (2003) have created a software package that identifies nonidentifiable parameters using a symbolic algebra package following earlier work by Catchpole and Morgan (1997) for the CJS model. Kendall and Nichols (2002) examined identifiability of the transition probabilities when some states are completely unobservable, such as nonbreeders.

Pradel et al. (2003) and Wintrebert (1998) have devised goodness-of-fit tests for the multistate model that are similar to those available for the CJS model and have prepared a computer program (UCARE) to perform these tests. One set of contingency tables extends the Test3 contingency tables for the CJS models and compares subsequent recaptures of subgroups of animals that are all seen and released at stratum s at time i. A new contingency table (TestM) tests if animals are mixing by comparing previous sightings for animals seen at stratum s at time j. Finally, a third set of contingency tables tests the memoryless property. Because of the large number of possible states, the number of counts in each cell of the contingency tables may be small, so rules are suggested for extensive pooling.

In cases where k and S are small, another possibility is to enumerate all the possible capture histories, compute the expected values under the final model, and perform a classical chi-squared goodness-of-fit test (Hestbeck et al. 1991). This approach is impractical in larger problems because the number of possible histories grows very rapidly and the data are then very sparse. Given these problems, a step-up approach starting from a simple model and moving to a more complex model is a good strategy.

Following Lebreton et al. (1992), the above modeling process can be generalized to multiple groups. For example, males and females may be assured to have separate movement, survival, and capture probabilities. The data demands for the most general model allowing for group, time, and stratum specific estimates are considerable.

Versions of the CJS model with survival or capture rates constant across groups or time have proved useful for single-stratum studies and the motivation to use simpler models is even greater in multistate models because of the large number of parameters. Using a generalization of the notation of Lebreton et al. (1992) and of Lebreton et al. (1999), a collection of models can be considered. For example, using a similar notation as in CJS models, the model $\{p_{g^*s^*t}, \phi_{g^*s^*t}, \psi_{g^*s^*t}\}$ has capture probabilities that are group, stratum, and time specific; survival probabilities that are group, stratum, and time specific; and movement probabilities that are group, stratum, and time specific. This is the most general model possible. Two examples of simpler models are the model $\{p_{g^*s^*t}, \phi_{g^*s^*t}, \psi_{g^*s}\}$ with capture probabilities that are group, stratum, and time specific; survival probabilities that are group, stratum, and time specific; and movement probabilities that are group and stratum specific but constant over time. The model $\{p_{g^*s^*t}, \phi, \psi\}$ has capture probabilities that are group, stratum, and time specific; survival probabilities that are constant over group, time, and stratum; and stratum movement probabilities that are constant over group, time, and stratum.

Model selection among the set of candidate models fit to the data can be based on the Akaikie Information Criterion (AIC) as described in chapter 1. Then the model with the lowest AIC represents the best choice in the trade-off between goodness of fit (more parameters) and parsimony (fewer parameters). Estimates from models with similar AIC can also be used in the model averaging procedures as described by Burnham and Anderson (1998). The greatest difficulties that will likely be encountered in the modeling exercise are determining the number of identifiable and estimable parameters and dealing with the effects of estimates on the boundary of the allowable range of values. Automatic parameter counting methods are under development (Reboulet et al. 1999; Catchpole and Morgan 1997) but these can be unreliable with sparse data. There are no simple ways to deal with parameter estimates on the boundary, because standard likelihood theory is no longer applicable.

Dupuis (1995) and Dupuis (2002) have developed Bayesian formulations of the Arnason-Schwarz model. These are likely to be too computationally intense for many problems, but provide a nice way of dealing with some of the technical problems encountered using standard likelihood methods.

Finally, because the number of parameters increases as the square of the number of strata, models with fewer strata can lead to more precise estimates from the same set of data. It is tempting to try and use the model selection methods to choose the number of strata. However, this must be done by constraining parameters in a general model rather than simply pooling strata. The latter involves comparing different models on different data sets (despite being derived from the same data), and may give misleading results.

Example—Movement of Canada Geese

Hestbeck et al. (1991) considered a study with multiple-recapture data of Canada Geese (*Branta canadensis*), which was further analyzed in Brownie et al. (1993). The geese were captured in three geographic strata (the mid-Atlantic, the Chesapeake, and the Carolinas) from 1984 to 1989. Only the 1986 to 1989 data will be used in this example. The capture histories are shown in table 8.1.

This dataset has only a single group so the group subscript *g* will be dropped in the remainder of this section. The most general model, model $\{\phi_{s*t}, \psi_{s*t}, p_{s*t}\}$ was initially considered where all parameters vary with stratum and time. In this model, the final capture probabilities are confounded with the final survival probability and so cannot be separately estimated. The former were therefore arbitrarily constrained to be 1.0. The estimates from this general model are shown in table 8.2. They seem to indicate an overall yearly survival rate of about 65–70% that may be constant over time, but varies among strata. The capture rates seem to vary among strata and possibly over time. The migration rates show strong fidelity to the area captured and appear to be roughly constant in the first two years of the study but with a large change in year 3 for stratum 3.

Based on the preliminary estimates, a series of models were considered, as summarized in table 8.3. The most appropriate model is model $\{\phi_s, \psi_{s*(t_1 = t_2, t_3)}, p_{s*(t_1 = t_2, t_3)}\}$, where the survival rates were constrained to be constant over time (but vary among strata), and the migration and capture rates were constrained to be equal in the first two years, with possible changes in the third year. This model has a high AIC weight. Estimates from the final model are shown in table 8.4. Some other simpler models may also be appropriate, such as one with the survival probabilities being equal for all strata, but further models were not considered. The estimates have remarkably small standard errors, but because the study observed over 10,000 birds this is not surprising.

Brownie et al. (1993) also fitted a non-Markovian model to the data and showed that there was evidence that the models considered here may not be fully appropriate. Their paper should be consulted for more details.

Other Uses for Multistate Models

Although the Arnason-Schwarz multistrata model was originally developed in the context of physical movement, the multistate model can also be used in a number of other contexts.

TABLE 8.1

Raw data (1986–1989) for the Canada goose study (provided by J. Hines, Patuxent Wildlife Center)

History	Count	History	Count	History	Count	History	Count	History	Count	History	Count
0001	158	0200	1040	0330	86	1200	81	2101	6	3002	17
0002	352	0201	27	0331	1	1201	9	2102	2	3003	21
0003	271	0202	117	0332	12	1202	12	2110	9	3010	9
0010	317	0203	11	0333	31	1210	11	2111	1	3012	1
0011	62	0210	59	1000	1135	1211	5	2112	4	3020	19
0012	31	0211	2	1001	40	1212	4	2120	7	3022	2
0020	748	0212	12	1002	39	1220	22	2121	1	3023	1
0021	56	0220	285	1003	2	1221	6	2122	4	3030	20
0022	234	0221	11	1010	63	1222	4	2131	1	3032	1
0023	7	0222	100	1011	22	1230	1	2200	423	3033	7
0030	504	0223	2	1012	8	1300	3	2201	10	3100	6
0031	14	0230	10	1020	68	1310	1	2202	45	3200	22
0032	71	0231	1	1021	8	2000	1559	2210	5	3202	3
0033	72	0232	5	1022	13	2001	18	2211	5	3220	5
0100	643	0233	2	1031	1	2002	85	2212	4	3221	1
0101	57	0300	694	1032	2	2003	4	2220	121	3222	1
0102	36	0301	7	1033	1	2010	33	2221	2	3230	2
0110	98	0302	66	1100	201	2011	6	2222	48	3233	1
0111	42	0303	60	1101	25	2012	3	2230	1	3300	57
0112	10	0310	13	1102	11	2020	163	2232	6	3301	2
0120	70	0311	1	1110	56	2021	3	2300	1	3302	2
0121	13	0313	2	1111	38	2022	51	2320	16	3303	15
0122	25	0320	59	1112	1	2023	1	2321	4	3320	3
0123	1	0321	2	1120	21	2030	7	2330	1	3330	10
0130	1	0322	13	1121	4	2032	1	3000	473	3332	3
0132	2	0323	2	1122	1	2100	55	3001	1	3333	13

Note. The three strata are 1, mid-Atlantic; 2, Chesapeake; 3, Carolinas.

TABLE 8.2

Estimates from the goose data from the model $\{\phi_{s*t}, \psi_{s*t}, p_{s*t}\}$

Parameter	Year	Stratum	Estimate	SE	Parameter	Year	Stratum	Estimate	SE
Recapture (p_i^s)	1987	1	0.423	0.03	Movement (ψ_i^{st})	1986–87	$1 \to 1$	0.703	0.024
		2	0.444	0.02			$1 \to 2$	0.286	0.024
		3	0.325	0.04			$1 \to 3$	0.010	0.005
	1988	1	0.426	0.03			$2 \to 1$	0.114	0.013
		2	0.473	0.02			$2 \to 2$	0.853	0.013
		3	0.252	0.02			$2 \to 3$	0.034	0.007
	1989	1	1.000[a]	—			$3 \to 1$	0.036	0.014
		2	1.000[a]	—			$3 \to 2$	0.180	0.030
		3	1.000[a]	—			$3 \to 3$	0.783	0.030
Survival (ϕ_i^s)	1986–87	1	0.634	0.02		1987–88	$1 \to 1$	0.669	0.024
		2	0.653	0.02			$1 \to 2$	0.319	0.027
		3	0.588	0.04			$1 \to 3$	0.011	0.005
	1987–88	1	0.632	0.03			$2 \to 1$	0.138	0.015
		2	0.634	0.02			$2 \to 2$	0.805	0.014
		3	0.702	0.04			$2 \to 3$	0.056	0.010
	1988–89	1	0.281[a]	0.02			$3 \to 1$	0.046	0.011
		2	0.279[a]	0.01			$3 \to 2$	0.201	0.023
		3	0.273[a]	0.02			$3 \to 3$	0.753	0.023
						1988–89	$1 \to 1$	0.714	0.025
							$1 \to 2$	0.281	0.025
							$1 \to 3$	0.005	0.004
							$2 \to 1$	0.159	0.013
							$2 \to 2$	0.822	0.013
							$2 \to 3$	0.019	0.005
							$3 \to 1$	0.056	0.012
							$3 \to 2$	0.396	0.026
							$3 \to 3$	0.547	0.027

[a] These parameters were confounded and only the products can be estimated.

TABLE 8.3
Summary of model fitting to the Canada goose data

Model	AIC_c	ΔAIC_c	AIC_c weights	Number of parameters
$\{\phi_s, \psi_{s^*(t_1 = t_2, t_3)}, p_{s^*(t_1 = t_2, t_3)}\}$	31661.213	0.000	0.85078	21
$\{\phi_s, \psi_{s^*(t_1 = t_2, t_3)}, p_{s^*t}\}$	31665.407	4.194	0.10450	24
$\{\phi_s, \psi_{s^*t}, p_{s^*t}\}$	31668.196	6.983	0.02591	30
$\{\phi_{s^*t}, \psi_{s^*t}, p_{s^*t}\}$	31670.307	9.094	0.00902	33
$\{\phi_s, \psi_{s^*t}, p_s\}$	31675.203	13.990	0.00078	24
$\{\phi_s, \psi_s, p_{s^*t}\}$	31689.268	28.055	0.00000	18

Live and Dead Recoveries

Lebreton et al. (1999) noted that the ordinary CJS model can be recast into a multistate model, with the two strata corresponding to live and dead. In this case, recaptures are possible only from live animals, and movement between live and dead states is unidirectional. In a similar fashion, the band-recovery models of Brownie et al. (1985) can also be cast into multistate framework, where the three strata correspond to live, newly dead, and previously dead. Only a single recapture is possible when the animal is recovered dead; and movement is again unidirectional. Models with both live recaptures and releases and dead recoveries (chapter 7) are also easily constructed. Now recaptures occur locally on a smaller study area while recoveries of newly dead animals can occur on a larger geographical scale. Lebreton et al. (1999) also show how to model radio-tracking data where dead animals are detectable from their functioning radio collars.

Local Recruitment of Breeders

The age of first reproduction is important in population regulation and life-history theory. Clobert et al. (1994) developed a model for mark–recapture data where animals are marked as young, but are not resightable until they commence to breed. Schwarz and Arnason (2000) considered the same problem and recast it as a Jolly-Seber experiment. Pradel and Lebreton (1999) examined the problem from a multistate perspective. Here, the two states are defined as nonbreeder or breeder. All animals are tagged at age 0 as nonbreeders, both breeders and nonbreeders share the same survival probabilities, initial breeding depends only on age, and experienced breeders breed systematically (i.e., do not skip a breeding

TABLE 8.4

Estimates from goose data from the final model $\{\phi_s, \psi_{s*(t_1=t_2,t_3)}, p_{s*(t_1=t_2,t_3)}\}$

Parameter	Year	Stratum	Estimate	SE
Recapture (p_j^s)	1987	1	0.426	0.020
		2	0.457	0.014
		3	0.280	0.020
	1988	1	0.426	0.020
		2	0.457	0.014
		3	0.280	0.020
	1989	1	0.451	0.043
		2	0.423	0.024
		3	0.446	0.064
Survival (ϕ_j^s)	1986–87	1	0.633	0.015
		2	0.647	0.012
		3	0.646	0.023
	1987–88	1	0.633	0.015
		2	0.647	0.012
		3	0.646	0.023
	1988–89	1	0.633	0.015
		2	0.647	0.012
		3	0.646	0.023

Parameter	Year	Stratum	Estimate	SE
Movement (ψ_j^{st})	1986–87	1 → 1	0.686	0.017
		1 → 2	0.303	0.017
		1 → 3	0.011	0.004
		2 → 1	0.125	0.009
		2 → 2	0.830	0.009
		2 → 3	0.044	0.006
		3 → 1	0.043	0.008
		3 → 2	0.200	0.018
		3 → 3	0.757	0.018
	1987–88	1 → 1	0.686	0.017
		1 → 2	0.303	0.017
		1 → 3	0.011	0.004
		2 → 1	0.125	0.009
		2 → 2	0.830	0.009
		2 → 3	0.044	0.006
		3 → 1	0.043	0.008
		3 → 2	0.200	0.018
		3 → 3	0.757	0.018
	1988–89	1 → 1	0.702	0.038
		1 → 2	0.293	0.038
		1 → 3	0.005	0.004
		2 → 1	0.152	0.021
		2 → 2	0.829	0.021
		2 → 3	0.019	0.006
		3 → 1	0.055	0.014
		3 → 2	0.406	0.047
		3 → 3	0.539	0.047

season). Lebreton et al. (2003) have extended this approach to dispersal–recruitment over several sites. Kendall et al. (2003) used multistate models along with Pollock's robust design to adjust estimates of recruitment for misclassification of the animal's state with an application to estimate the change in breeding status for Florida manatees (*Trichechusmanatus latirostris*).

Individual Covariates That Change over Time

Nichols et al. (1994) used multistate models where states were defined by individual covariates (weight classes of meadow voles) that changed over time, rather than geographical sites. This is very general, and any continuous covariate could be stratified. One area that requires further research is the choice of class boundaries and the comparison of models with different class boundaries. States can also be defined based on discrete covariates such as reproductive success, e.g., breeders vs. nonbreeders. These have been used to study trade-offs between different reproductive strategies (Nichols et al. 1994).

8.3 The Jolly-Seber Approach

Despite the focus of current capture–recapture experiments and models for estimating survival and movement parameter, the original motivation for these methods was on estimating abundance. For example, Jolly (1965) and Seber (1965) were primarily interested in estimating the number of animals alive at each sampling occasion. The general class of models dealing with abundance and survival (JS models) is a generalization of the CJS models that estimated only survival and capture probability.

Far less work has been done on multistate models that include estimating abundance, for several reasons. First, the experimental protocol necessary to estimate abundance is more rigorous than that to estimate survival or movement. In the latter cases, the process by which animals are first captured and tagged is relatively unimportant, as long as the subsequent survival and movement of the marked animals is representative of the larger population. In contrast, estimating abundance requires that the process by which new animals are captured gives a random selection from the entire population. As well, the definition of the relevant population is also important. If captures take place at a single geographical location of a highly mobile animal, then it is not clear whether the relevant population is defined as those within 1, 5, or 20 km of the sampling site.

Second, heterogeneity in capture probabilities can lead to biases when estimating abundance. At the extreme, animals with very low capture probabilities may be missed entirely in the estimates. Recent work in the CJS framework of trying to estimate population changes rather than abundance looks promising, and Schwarz (2001) has shown that that heterogeneity in capture probabilities causes little bias in estimates of population change in the simple JS model.

Third, temporary emigration also leads to severe biases in the estimates and makes it difficult to clearly define the relevant population in terms of whether the estimate of population size should include temporary migrants or exclude them. Kendall et al. (1995, 1997), Schwarz and Stobo (1999), and others have shown how to use Pollock's (1982) robust design to estimate both components of the population.

The latter two concerns have been the primary motivation of work in estimating abundance using stratified methods. In order of prevalence of use, these can be broadly classified as

- two-sample experiments to estimate abundance, where the focus has been on reducing the biases in the simple Petersen estimator caused by heterogeneity in capture probabilities;
- multisample closed populations, where the focus has been on modeling movement to reducing biases caused by heterogeneity of capture and temporary absences from a single sampling location; and
- multisample open populations, where the focus has been modeling movement to reduce biases caused by heterogeneity of capture and temporary absences from a single sampling location.

Two-sample Stratified Experiments

In a two-sample experiment, animals are captured, tagged, and released in the first sample and allowed to mix with the remaining population. Then a second sample is taken and inspected. The simplest estimator of abundance is then the Lincoln-Petersen estimator

$$\hat{N}_{\text{Petersen}} = \frac{n_1 n_2}{m_2}$$

where n_1 is the number of animals seen at time 1, n_2 is the number of animals seen at time 2, and m_2 is the number of animals seen at times 1 and 2. Small-sample, bias-corrected versions are also available.

The Lincoln-Petersen estimator is a consistent estimator of the population size when either or both of the samples are a simple random sample so that either all animals in the population have the same probability of being tagged or all animals have the same probability of being captured in the second sample. It is also assumed that the population is closed, there is no tag loss, the tagging status of each animal is determined without error, and tagging has no effect on the subsequent behavior of the animals (chapter 2).

There are many cases where the capture probabilities are not homogeneous over the animals in either sample. For example, in surveys of spawning populations, fish are caught and tagged as they pass a common point prior to the spawning grounds, over the course of several days or weeks. It is unlikely that the tagged fish are a simple random sample. This would require, at a minimum, that a constant proportion of the fish be tagged each day. Similarly, the recaptures may be obtained by searching carcasses on selected spawning area. The recapture probabilities are likely to vary by recovery area and over time, and some spawning areas may not be sampled at all. Lastly, it is unlikely that fish passing the tagging site early in the run mix completely with fish that pass the tagging site late in the run. Under these circumstances, the simple Petersen estimator formed by pooling over time and space all of the tags applied and all of the recoveries may be biased and this bias can be considerable (Arnason et al. 1996b). Schwarz and Taylor (1998) derived an approximation to the size of the bias and showed that it was related to the correlation between the two capture probabilities. Substantial bias is likely to occur if there is great variability in both the capture and recapture probabilities (e.g., through sex or size effects), or if there is little mixing of the animals as they move from the tagging to the recovery locations.

One method of reducing this bias is to stratify by time, space, or other characteristic. The captures at time 1 are placed into S nonoverlapping strata while captures at time 2 are placed into T nonoverlapping strata at the time of recovery. Unlike the AS models, the two-sample stratified experiment allows for different number of strata at each sample time. For example, in geographic stratification, tagging strata may be different stocks of fish caught on different breeding areas, and recovery strata may be different fishing areas. In temporal stratification, tagging and recovery may take place over several weeks, and individual strata correspond to time intervals (e.g., weeks). Stratification can also be by permanent attributes (e.g., sex) or changing attributes (weight class).

Capture histories can again be constructed to represent the data as in the previous section. The history vectors take the forms $\{s, t\}$, $s = 1, \ldots, S$, $t = 1, \ldots, T$ for animals captured in stratum s at time 1 and in stratum t at time 2; $\{0, t\}$, $t = 1, \ldots, T$ for animals not captured at time 1 but captured in stratum t at time 2; and $\{s, 0\}$, $s = 1, \ldots, S$ for animals captured in stratum s at time 1 and not captured at time 2. The number of animals with these histories is often placed in a rectangular array as shown in table 8.5.

TABLE 8.5

Summary statistics for female pink salmon tagged at Glen Valley and recovered in the Fraser River Main Stem arranged in a rectangular array

Release stratum[a]	Fish tagged	Recovery stratum[b] Fraser River Main Stem					Not seen again (n_{s0})
		8 Oct 91	14 Oct 91	20 Oct 91	26 Oct 91	1 Nov 91	
10 Sep 91	2881	13	2	2	1	0	2863
15 Sep 91	5478	28	14	10	6	2	5418
20 Sep 91	4661	13	18	12	1	0	4617
25 Sep 91	5978	8	10	12	5	0	5943
30 Sep 91	2386	0	4	1	1	1	2379
5 Oct 91	507	0	0	0	0	0	507
Untagged (n_{0t})		18,484	15,317	15,064	9355	993	
Total recoveries		18,546	15,365	15,101	9369	996	

[a] Each release stratum consists of five days of releases beginning with the indicated date.

[b] Each recovery stratum consists of six days of pitching beginning with the indicated date in each of the spawning areas.

The parameters of the model are the capture probabilities at times 1 and 2, p_1^s, $s = 1, \ldots, S$ and p_2^t, $t = 1, \ldots, T$; the movement parameters among strata between times 1 and 2, ψ^{st}, $s = 1, \ldots, S$, $t = 1, \ldots, T$; and the population sizes in the strata at time 1 and time 2, N_1^s and N_2^t. The model allows "leakage" between the two sampling occasions, i.e., $\sum_{t=1}^{T} \psi^{st} \leq 1$, which could be caused by mortality or movement out of the study area. This is unlike the AS models where the movement probabilities must sum to 1.

The usual assumptions for capture–recapture experiments are made and it is additionally assumed that

    1. animals behave independently of one another in regard to moving among strata;

    2. all tagged animals released in a stratum have the same probability distribution of movement to the recovery strata;

    3. all animals in a recovery stratum behave independently with regard to being caught and all have the same probability of being caught;

    4. no tags are lost; and

    5. all animals that are recovered are correctly identified as to the tagging status, and if tagged, the tag number is correctly recorded, although Schwarz et al. (1999) extend this situation to cases where some of the tags in the recovery strata are counted but not read to identify the stratum of release.

In addition, one or both of the following assumptions is usually made depending whether the goal of the study is to estimate the number of animals in the tagging or recovery strata:

    6a. the movement pattern, death, and migration rates are the same for tagged and untagged animals in each tagging stratum (required to estimate the total number of animals in the tagging strata); or

    6b. the population is closed with respect to movement among strata (required to estimate the total number of animals at the recovery strata).

Model Fitting and Estimation

In the most straightforward case there is a one-to-one correspondence between the tagging and recovery strata. This will occur, for example, with stratification based on permanent attributes (e.g., sex) where animals cannot move from one stratum to another. In these cases, the problem reduces to s independent Petersen estimates. The total abundance is found by summing these independent estimates and the overall standard error is easily found.

Three common estimators are used in the case of the more general stratified samples. The pooled Petersen is the simplest estimator and is formed by ignoring the stratification completely and computing a simple Petersen estimate based on the total releases and total recaptures:

$$\hat{N}_{pooled} = \frac{(n_{1+} + n_{2+} + \cdots + n_{S+})(n_{+1} + n_{+2} + \cdots + n_{+T})}{n_{11} + n_{12} + \cdots + n_{ST}}$$

where $n_{s+} = n_{s0} + n_{s1} + n_{s2} + \cdots + n_{sT}$ and $n_{+t} = n_{0t} + n_{1t} + n_{2t} + \cdots + n_{St}$.

Darroch (1961) and Seber (1982), summarized three common circumstances under which the pooled-Petersen remains a consistent estimator:

    1. the tagging probabilities are the same in all release strata;

    2. the recovery probabilities are the same in all recovery strata, and the same degree of closure of movement occurs for each release stratum, i.e., the same fraction of each release stratum is subject to recovery in the recovery strata; and

    3. there is complete mixing of animals from all the release strata before animals are recovered, and the movement pattern for tagged and untagged animals is the same.

Darroch (1961) discussed formal statistical tests that can be used to examine whether the pooled-Petersen estimate is consistent. In particular, he discussed two simple contingency table methods that have been implemented by Arnason et al. (1996a).

Schaefer (1951) developed an estimator that has been rediscovered numerous times (e.g., Macdonald and Smith 1980; Warren and Dempson 1995). This takes the form

$$\hat{N}_{\text{Schaefer}} = \sum_{s=1}^{S} \sum_{t=1}^{T} \frac{n_{s+}n_{+t}n_{st}}{(n_{s1} + n_{s2} + \cdots + n_{sT})(n_{1t} + n_{2t} + \cdots + n_{St})}$$

Surprisingly, although the Schaefer estimator has been used in the literature in three common disguises, its properties were not systematically investigated until recently (Schwarz et al. 2002). In fact, the Schaefer estimator has essentially the same performance (in terms of bias and standard error) and requires the same conditions for consistency as the pooled-Petersen estimator. Consequently, there is no advantage to using the Schaefer estimator over the pooled-Petersen estimator, and it is no longer recommended for use.

Lastly, a Petersen type of estimator can be constructed that accounts for the stratified nature of the experiment and models movement. As in the AS model, the probability of each history vector must be expressed in terms of the parameters. However, it is more convenient to express the expected counts in terms of the parameters to give

$$E(n_{\{s,t\}}) = N_1^s p_1^s \psi^{st} p_2^t$$

$$E(n_{\{s,0\}}) = \sum_{t=1}^{T} N_1^s p_1^s \psi^{st} (1 - p_2^t)$$

and

$$E(n_{\{0,t\}}) = \sum_{s=1}^{S} N_1^s (1 - p_1^s) \psi^{st} p_2^t$$

A likelihood function can be constructed based on these expected counts, assuming a multinomial model from each release. Plante et al. (1998) and Banneheka et al. (1997) show that all of the parameters cannot be individually estimated. However, certain functions of the parameters may be estimated under each of two different scenarios.

If the number of tagging strata is less than or equal to the number of recovery strata ($S \leq T$), and if the assumption of the same movement patterns for tagged and untagged animals holds, but not necessarily closure, then Plante et al. (1998) and Banneheka et al. (1997) showed that the number of animals in the population at the time of tagging can be estimated. If the number of tagging strata is greater than or equal to the number of recovery strata $(S \geq T)$, and there is closure of the population with respect to movement to the recovery strata, then the population size at the time of recovery can be estimated. When $S = T$ and both assumptions 6a and 6b above are satisfied, then the population sizes at the time of tagging and at the time of recovery are identical, and both can be estimated. The case of $S = T$ was fully developed by Darroch (1961) and summarized by Seber (1982).

For most experiments, the assumptions of the same movement patterns for tagged and untagged animals, but not necessarily closure over recovery strata, may be more tenable than the assumption of closure with respect to movement among strata. For example, in a typical spawning escapement survey, the population is not closed because there is mortality between the tagging site and the spawning areas, fish leave the population because of spawning in other than the recovery sites, fish may arrive and spawn before or after recovery takes place, and fish may be removed through a fishery that takes place during the run. For this reason, the stratified-Petersen analysis with $S \leq T$ is the most common to estimate the number of animals in the population at the first sampling occasion.

The usual goodness-of-fit test comparing the observed and expected counts for each capture history can be used to assess the fit of a model.

Plante et al. (1998) developed general maximum likelihood theory for the cases where the number of release strata did not match the number of recapture strata. Banneheka et al. (1997) developed a least-squares estimator and found its properties. Arnason et al. (1996a, b) examined conditions under which stratified estimators were superior to the usual Petersen estimator, and have developed a user-friendly computer package that implements all of these recent advances. Finally, Schwarz and Taylor (1998) provided a comprehensive development and review of the use of the stratified-Petersen estimator, with an application to estimating the number of salmon returning to spawn.

Fitting models like these to real data is challenging because of two common problems:

1. values of n_{st} close to or equal to zero; and
2. rows of the $\{n_{st}\}$ matrix that are near or exact multiples of each other or, more generally, rows that are exactly or nearly linearly dependent (i.e., some linear combination of the rows sum to zero).

Elements of n_{st} that are exactly equal to 0 are a problem because sampling zeros (values of $n_{st} = 0$ that occur just by chance) cannot be distinguished from structural zeros (physically impossible movements, such as recoveries occurring before releases in time stratification). In both cases, the maximum likelihood estimate of the movement probability is zero. This is potentially a problem for sampling zeros because there was some movement of the population that, just by chance, was not detected.

Linear dependency among the rows of the $\{n_{st}\}$ matrix is a problem because the resulting estimators are no longer consistent (Darroch 1961; Banneheka et al. 1997), although linear dependency among the columns of the matrix is not a problem as long as the rank of the matrix is equal to the number of rows. In all cases, the stratified data may have to be modified by either dropping rows or columns, or by pooling two or more rows or columns, and then refitting the model with the new stratification. Schwarz and Taylor (1998) examine in detail the effects of pooling or dropping both rows and columns, and the resulting bias in the estimates. There is not as yet any objective way to determine the optimal pooling of rows or columns.

Example

Fraser River pink salmon (*Oncorhynchus gorbuscha*) swim up the Fraser River in mid-September to spawn in more than 60 different tributary streams. Here we describe a tag-recovery study where approximately 22,000 female pink salmon were tagged and released between 10 Sep 91 and 8 Oct 91 at Duncan Bar (Glen Valley) downstream of the spawning areas. Tag recovery was done by searching spent carcasses (called dead pitching) found on the stream banks after spawning. Approximately 300,000 female carcasses were dead-pitched from 2 Oct to 12 Nov 1991.

Schwarz and Taylor (1998) examined the recoveries for both sexes from five spawning areas. In this example data on the female releases at Glen Valley and recoveries on the Fraser River Main Stem spawning areas are examined in order to estimate the number of fish passing the tagging site.

The data were too sparse to stratify on a daily basis. Schwarz and Taylor (1998) showed that pooling of recovery strata may be done in a fairly arbitrary fashion without affecting the consistency of the estimates. They found it necessary to stratify releases into 5-day intervals (10–14 Sep 1991, 15–19 Sep 1991, etc.) and recoveries into 6-day intervals (8–13 Oct 1991, 14–19 Oct 1991, etc.). The resulting 6×5 matrix of recoveries on the Fraser River Main Stem with summary statistics is shown in table 8.5.

The estimate from the pooled-Petersen method is 7.70 million fish, with a standard error of 0.59 million. However, the contingency table tests for the consistency of the pooled-Petersen method showed some evidence that complete pooling may not be appropriate.

Before, using the stratified-Petersen, it was noticed that no recoveries were obtained from the release group dated 5 Oct 1991 and so this row was deleted. Because of the small number of fish tagged in this release group, these observed zeroes may not be structural zeroes. Fish from this release group may spawn in the Fraser Main Stem, but there may have been too few tagged to detect such spawning. As well, it was necessary to pool the first two and last three rows of the data to obtain nonnegative estimates.

Final results are shown in table 8.6. A chi-squared goodness-of-fit test of the stratified-Petersen model to the pooled data shows some evidence of a lack of fit in two cells, but given the relatively small counts the lack-of-fit is assessed not to be serious. The estimated number of fish to pass the tagging sites in the first five release periods is 8.35 million with an standard error of 0.82 million fish. The pooled-Petersen estimator is 8% less than the stratified estimate.

Discussion

The stratified-Petersen estimator should be used when estimates of the individual strata population sizes are of direct interest. Even if the individual strata are not of interest, it can significantly reduce the bias of a pooled-Petersen estimator when there is substantial variation in and correlation among the initial-capture and final-recapture probabilities. In many escapement studies, the natural pattern of returns induces a positive correlation between these two probabilities and this results in a negative bias in the pooled-Petersen and Schaefer estimates.

However, numerous practical problems often make it difficult to use the stratified-Petersen estimator:

- Recovery data are relatively sparse. Even with the application of over 40,000 tags and pitching over half a million carcasses in the salmon example, the number of tags returned was often small after stratification by sex, time of tagging, spawning ground, and time of recovery.
- Release strata are not completely distinct. In many cases, migration patterns are slowly changing over time and a release stratum consists of fish with migration patterns from the previous and the next stratum. This causes near linear dependencies to exist in the observed and expected recovery matrices.
- The animals from different release strata appear to mix on their way to the recovery strata. This implies that the recovery matrices will not have an upper-triangular structure and allowed linear dependencies to occur.

TABLE 8.6

Detailed results from analyzing female Glen Valley releases recovered in the Fraser Main Stem only—final pooling

Stratum	Fish released	Observed recovery statistics with fitted values beneath					Not seen again	$\hat{N}_i^s$ (millions)	SE $\hat{N}_i^s$ (millions)
		8 Oct 91	14 Oct 91	20 Oct 91	26 Oct 91	1 Nov 91			
1 & 2	8,359	41	16	12	7	12	8,281	1.57	.84
10 Sep 91 & 15 Sep 91		40.0	14.4	12.0	8.6	2.1	8,281		
3 & 4 & 5	13,025	21	32	25	7	1	12,939	6.78	1.37
20 Sep 91 & 25 Sep 91 & 30 Sep 91		20.0	24.3	24.7	14.9	1.2	12,939		
Untagged		18,484	15,317	15,064	9355	993	Total pop:	8.35	.82
		18,484.2	15,326.3	15,064.3	9345.5	992.7			

Note. Each entry has the observed and expected counts used in the goodness-of-fit test. χ^2 goodness-of-fit test: $\chi^2 = 7.1$, 3 df, p value = 0.068. Pooled Petersen estimate: 7.70 million with an estimated SE of 0.59 million.

In many cases, extensive pooling will be required to obtain admissible estimates, which partially defeats the purpose of stratification.

The pooled-Petersen and stratified-Petersen provide methods at opposite ends of the spectrum with regards to assumptions about capture and movement. The pooled-Petersen assumes that all animals have identical capture and movement probabilities, while the stratified-Petersen lets each release stratum have its own distinct movement pattern. While this may be realistic for geographical stratification, it is, paradoxically, too flexible for temporal stratification. As was seen with the pink-salmon data, migration patterns are likely to change slowly over time, leading to situations where near linear dependencies can occur. Intermediate models have been developed. For example, Schwarz and Dempson (1994) model the movement process of tagged outgoing salmon smolt. This leads to a dramatic decrease in the number of parameters, fewer problem with sparse data, and very precise estimates.

8.4 Multisample Stratified Closed Populations

Heterogeneity in capture probabilities is a well-known problem in closed-population models. The Otis et al. (1978) suite of models (M_0, M_t, M_b, M_h, . . .) allow for heterogeneity in capture probabilities caused by time, behavior, or intrinsic animal characteristics and much modeling effort has been devoted to these models (chapter 2). The motivation for considering multisite, closed-population models is to account for heterogeneity among animals that can be identified with strata. These strata can be quite general (e.g., weight class, geographical location, type of trap).

The survey protocol is similar to that of the AS model. At each of k sampling times, animals are captured, tagged if necessary, and released. At each capture occasion, the stratum of capture ($s = 1, . . . , S$) is also recorded. As noted earlier, the key difference in the survey protocols when estimating abundance is that the process by which animals are first tagged must be carefully designed. A key assumption is that untagged animals are a simple random sample from the untagged animals in the population with the same capture probabilities as tagged animals. This experiment also differs from the two-sample case in that the number of strata (S) is the same at all sampling times.

The raw data from the experiment consist of a history vector of length k for each of the captured animals in the same format as in the AS model of section 8.1. The key difference in interpreting the history vector from the interpretation in section 8.1 is that the animal is known to be alive both prior to its initial capture and following its last capture.

The parameters of this model are the total population size over all strata (N), the initial distribution of these animals among the strata (χ^s), the capture probabilities at each time point (p_j^s), the movement probabilities between sampling times j and $j+1$ (ψ_j^{st}), and possibly loss-on-capture parameters. The latter three parameter sets have an identical interpretation to those of the AS model. Because the population is closed, there are no survival parameters.

In an experiment with k sample times and S strata, there are a total of $(k-1)(S^2 - S)$ movement parameters $\psi_j^{st}, j = 1, \ldots, k-1$ and $s = 1, \ldots, S$, but $\psi_j^{s1} + \psi_j^{s2} + \cdots + \psi_j^{sT} = 1$ for each s and j; kS capture parameters p_j^s, $j = 1, \ldots, k$ and $s = 1, \ldots, S$; kS loss-on-capture parameters; one population size parameter; and $S-1$ parameters for the initial distribution among the strata, with $\chi^1 + \chi^2 + \cdots + \chi^s = 1$.

In a similar fashion to section 8.1, the parameters can be used to model the probability of a particular capture history. However, it is more instructive to write out the expected counts for each history, since this explicitly includes the abundance term. For example, assuming only 2 strata, the expected number of animals with the history vector $h = (1, 2, 0, 1)$ is

$$N\chi_1 p_1^1 \psi_1^{12} p_2^2 [\psi_2^{21}(1 - p_3^1)\psi_3^{11} + \psi_2^{22}(1 - p_3^2)\psi_3^{21}]p_4^1$$

where N is the initial population size, χ_1^1 is the probability that the animal started in stratum 1 at time 1, p_1^1 is the probability that the animal in stratum 1 at time 1 was captured, ψ_1^{12} is the probability that the animal released in stratum 1 at time 1 will be present in stratum 2 at time 2, p_2^2 is the probability that the animal in stratum 2 at time 2 will be captured,

$$[\psi_2^{21}(1 - p_3^1)\psi_3^{11} + \psi_2^{22}(1 - p_3^2)\psi_3^{21}]$$

is the overall probability that an animal present in stratum 2 at time $j = 2$ will be in stratum 1 at time 4 (a term that must account for the two possible paths from stratum 2 to stratum 1 between times 2 and 4), and p_4^1 is the probability that the animal stratum 1 at time 4 was captured.

Note that after the first capture of an animal, the subsequent modeling for the capture history is identical to the AS model. The complexities in writing out the expressions for each capture history are the same as in the AS model, but with the additional complexity that the all possible movements from time 1 to the first capture must be included because it is known that the animal is alive from the start of the experiment. As well, it is also possible to write out the probability for the animals never seen in the experiment.

The likelihood is constructed as the product of probabilities—one for each animal captured and one for the animals never captured but known to be alive. All parameters are identifiable, though some may be non-estimable in cases with sparse data. No closed-form expressions exist for the maximum likelihood estimators, but Schwarz and Ganter (1995) have written a SAS program to compute the values.

The model makes a number of assumptions in addition to those made for the CJS model:

- all data are recorded without error with respect to stratum of capture;
- there are homogeneous transition probabilities for all animals in stratum s at time j; and
- there is a lack of memory, i.e., the transition probability and recapture probabilities do not depend on the past history of the animal.

The models can also be generalized to multiple groups or simplified using models with survival or capture rates that are constant across groups or time. Notation for these models can be generalized from the previous chapter. Model selection among a set of candidate models for the data may be based on the Akaike information criterion (chapter 1).

Example

During the spring migration of a major population of barnacle geese (*Branta leucopsis*), the birds stage (to prepare for migration) in five areas of the Schleswig-Holstein region of northern Germany for several weeks. Numerous researchers have banded many geese on their breeding and wintering areas using colored leg bands that are visible during the staging period. There is little mortality during the staging period. Consequently, the population is nearly closed during this period (Ganter 1995).

From mid-January to early April, members of a research team visited the five areas. During each visit, team members recorded the band number of any banded goose spotted. For reasons outlined in Schwarz and Ganter (1995), the data were grouped into five two-week intervals with interval 1 starting 30 Jan 1990. The capture histories for this experiment are given in table 8.7.

Even after grouping, areas 1 and 5 had no sighting effort in some intervals. Hence, the full model cannot be used, and only models assuming constant transition probabilities over time can be considered (Schwarz and Ganter 1995). The model $\{p_{s*t}, \psi_s\}$ was considered for the complete data. Information on movement from area 5 is obtained from releases at time 2 while information on movement from area 1 is obtained from releases at times 3 and 5. A summary of the fitted model is shown in table 8.8. The

TABLE 8.7
Capture histories for the Barnacle goose experiment

History	Count	History	Count	History	Count
00002	3	00404	1	22023	1
00003	1	00440	4	22040	1
00010	1	00442	1	22210	1
00011	5	01002	1	22220	3
00020	7	01200	1	22222	1
00022	11	01202	1	22240	1
00030	1	01240	1	22300	1
00032	1	02000	2	22333	1
00033	2	02001	1	23233	1
00034	1	02004	1	30000	3
00040	1	02011	1	30002	1
00041	1	02020	4	30003	1
00044	9	02200	1	32001	1
00200	2	02202	1	32200	1
00202	2	02220	1	33000	1
00220	13	03003	1	33220	1
00222	4	03020	1	33333	1
00330	1	03300	2	40000	3
00400	1	03330	1	40002	1
00401	1	03333	2	40003	2
00402	1	04200	1	40022	1
00404	1	20000	4	40200	2
00440	4	20001	1	40400	2
00442	1	20002	2	41001	1
01002	1	20004	1	42004	1
01200	1	20020	1	42010	1
01202	1	20022	2	42040	1
01240	1	20200	1	50000	8
02000	2	20202	1	50001	1
00202	2	20220	3	50002	1
00220	13	20222	1	50200	1
00222	4	20230	1	53400	1
00330	1	20330	1		
00400	1	20332	1		
00401	1	22000	5		
00402	1	22020	6		

Note. Some of the zeros are structural zeros where no sighting effort was applied. Refer to Schwarz and Ganter (1995) for details.

TABLE 8.8

Results from multistate multisample closed-model fitting to all areas for the Barnacle goose data

Model	Log-likelihood[a]	Parameters[b]	AIC
$\{p_{s*t}, \psi_s\}$[c]	−856.6	43	1799.2
$\{p_{lin(visits)}, \psi_s\}$	−874.3	34	1816.6

[a] Ignores constant term in the log-likelihood.
[b] Ignores structural zeros.
[c] Goodness-of-fit test has $\chi^2 = 67.0$ with 113 df, p value = 0.99.

goodness-of-fit test failed to reject this model and residual plots also showed few problems. A simpler model where resighting rates were constant over time, $\{p_s, \psi_s\}$, was strongly rejected although results are not shown. This is again not surprising given the wide range in the number of visits made to the areas. Another simpler model, $\{p_{f(visits)}, \psi_s\}$, where the sighting probabilities increase with the number of visits made to an area, was also examined, but had little support. The final estimates are shown in table 8.9. These indicated a high fidelity to area 2 and 5, but substantial movement among areas 1-2, 3-2, 3-5, 4-2, and 4-5. The relatively strong fidelity to sites 2 and 5 may be due to the physical conditions of the sites as described in Ganter (1995).

Discussion

Many studies of animal movement have been fragmented (e.g., effort has not been expended simultaneously in all areas in all times). This implies that many general models cannot be fitted because of nonidentifiability of the parameters. However, in some cases, subsets of the data are more uniform. An analysis of a subset of the data may provide some information about likely models that may be appropriate for a population. Under a somewhat restrictive model of equal movement rates over time, even fragmented studies can yield important information.

Finally, as noted by Brownie et al. (1993), multisite models allow the modeling of heterogeneity in capture and survival rates. For example, model M_b of Otis et al. (1978) assumes that once an animal is captured, its subsequent capture probabilities remain fixed. An even richer class of models can be fitted if we allow the behavior modification to be probabilistic in nature by defining two strata corresponding to behaving as if not captured and behavior modified because of capture. Movement parameters can be defined to model the fraction of animals that forget they have been captured in each time period. This would allow the full capture

history of the animal to be used, rather than discarding all subsequent recaptures after the first as is done in model M_b of Otis et al. (1978).

8.5 Multisample Stratified Open Populations

This is a generalization of the Jolly-Seber model to the case of multistrata.
Again, the motivation for this model is the estimation of movement and to
account for heterogeneity in the capture probabilities among the different
strata. This was first considered in detail by Arnason (1972, 1973) who
developed moment estimates for the case of three sample times. Schwarz
et al. (1993b) developed moment estimates for the general k-sample case.
Surprisingly, very little additional theoretical work has been done on this
model, presumably because of the very exacting experimental protocols
required.

Fortunately, the results of earlier sections can be used for a large part of
the analysis. As in the ordinary Jolly-Seber model, the probability for
each capture history will consist of components describing the first capture (which is used to estimated abundance), and then the subsequent recapture history of marked animals. The latter will be identical to that in
the AS model of movement discussed in section 8.1. The first component
is likely to be very complicated and strategies similar to that used by
Schwarz and Arnason (1996) and Schwarz and Ganter (1995) will likely
be necessary to describe new animals entering the population using a
complex multinomial distribution.

However, based on experience with abundance models, a close approximation to the fully maximized likelihood can be obtained by fitting the
AS model of section 8.1 to the recapture data only. Then the abundance
at each time point can be estimated using the moment equations

$$\hat{N}_j^s = \frac{n_j^s}{\hat{p}_j^s}$$

where the subscripts refer to the estimate in stratum s at time j. As in the
simple Jolly-Seber model, estimates of recruitment to each stratum can
be obtained from the moment estimates

$$\hat{B}_j^s = \hat{N}_{j+1}^s - \hat{N}_j^s \hat{\phi}_j^s$$

where losses on capture have been ignored. As in the Jolly-Seber model,
this estimator may result in estimates of "births" less than zero, suggesting shortcomings in this approach to evaluate recruitment. Because of
the large data requirements and strict protocols required for abundance
estimation, there are few (if any) examples in the literature. Dupuis

TABLE 8.9

Estimates from model $\{p_{s \cdot t}, \psi_s\}$ (equal movement over time, unequal resighting rates over time) for the Barnacle goose example

Movement rates arranged in a matrix. The entry in row s and column t represents movement from stratum s to stratum t.

$$\hat{\psi} = \begin{bmatrix} .494 & .497 & .000 & .000 & .009 \\ .021 & .892 & .031 & .028 & .028 \\ .001 & .220 & .528 & .052 & .200 \\ .076 & .238 & .000 & .488 & .198 \\ .000 & .026 & .032 & .000 & .942 \end{bmatrix} \quad SE(\hat{\psi}) = \begin{bmatrix} .133 & .126 & .000^a & .000^a & .000^a \\ .011 & .038 & .011 & .011 & .032 \\ .002^a & .077 & .073 & .037 & .078 \\ .038 & .067 & .002^a & .046 & .083 \\ .000^a & .038 & .026 & .000^a & .042 \end{bmatrix}$$

Remaining parameters:

Parameter	Interval	Stratum	Estimate	SE
Initial distribution	1	1	.001	.011[a]
(ε^s)		2	.174	.048
		3	.086	.040
		4	.572	.186
		5	.167	.250
Resighting	1	1	.000[b]	—
(p_j^s)		2	.999	.000[a]
		3	.472	.156
		4	.108	.032
		5	.293	.506
	2	1	.484	.338
		2	.514	.076
		3	.999	.000[a]
		4	.013	.014
		5	.000[b]	—
	3	1	.000[b]	—
		2	.489	.064
		3	.999	.000[a]
		4	.339	.090
		5	.000[b]	—
	4	1	.682	.178
		2	.587	.068
		3	.997	.019[a]
		4	.876	.108
		5	.000[b]	—
	5	1	.999	.000[a]
		2	.305	.052
		3	.996	.000[a]
		4	.971	.002[a]
		5	.000[b]	—

[a] Standard errors for estimates close to 0 or 1 are difficult to determine and reported values are likely underestimates of actual precision.
[b] No sighting effort expended—a structural zero.

Figure 8.2. Radio-collared polar bear (*Ursus maritimus*) and new cub on the sea ice north of Kaktovik, Alaska, April 1986. (Photo by Steven C. Amstrup)

and Schwarz (in preparation) have developed a Bayesian approach to estimate abundance, survival, and movement of walleye in Mille Lac, Minnesota.

8.6 Chapter Summary

- Motivation is provided for considering models where animals are in several groups (such as geographical areas), and these groups change with time.
- The Arnason-Schwarz model that allows animals to move between states is reviewed, including the earlier work that led to this model, the data collection procedures, and model fitting and selection. The model is illustrated using some data on the captures and recaptures of Canada geese.
- Applications of the model are described with live and dead recoveries of animals, the recruitment of breeders, and covariates that change with time.
- The Arnason-Schwarz model does not allow the direct estimation of population sizes. Models of the Jolly-Seber type that do allow this estimation are discussed, starting with the two-sample, closed-

population case, which is illustrated by an example on the recaptures of male and female pink salmon from five spawning areas.

- The multisample closed-population case is discussed, with an example involving the movement of barnacle geese.
- There is a brief discussion of the multisample case for an open population. It is recommended that for the estimation of population sizes the Arnason-Schwarz model should be fitted to the recapture data, and then reciprocals of estimated capture probabilities can be used to estimate the sizes.

Nine

Examples

TRENT L. McDONALD, STEVEN C. AMSTRUP,
ERIC V. REGEHR, AND BRYAN F. J. MANLY

9.1 Introduction

In this chapter we provide empirical examples of how to use the models described in earlier chapters to analyze real-world data sets. With numbers rather than symbolic notation, we illustrate some practical aspects of capture–recapture modeling. We provide explicit examples and instructions on the mechanics of model building, and illustrate how to set up analyses in program MARK. We hope that the examples provided in this chapter will reduce the anxiety that often accompanies attempts at new analytical approaches.

Much of this chapter is built around capture–recapture data collected on the European dipper (*Cinclus cinclus*) (Marzolin 1988). This data set is typical of many capture–recapture studies, and has been analyzed a number of times in the literature (e.g., Lebreton et al. 1992). The dipper data provide a convenient framework for the application of several of the models covered in previous chapters of this book. We start by illustrating how to compute two early and relatively simple capture–recapture models for open populations (Jolly-Seber and Manly-Parr, chapter 3). We then work through a series of more complicated Cormack-Jolly-Seber (chapter 5) analyses of the dipper data using program MARK. We also present an alternative parameterization of the CJS model, in which explanatory covariates are framed in a general regression context. This general regression approach allows capture–recapture data to be analyzed in a familiar and intuitive framework, facilitates complicated covariate structures, and may be more easily understood by newcomers to capture– recapture analysis. Next, we treat the dipper data as if they were collected from a closed population and perform one of the recent approaches to closed-population assessment (i.e., the Huggins model, chapter 4). We then explain how to assess the goodness of fit of a CJS model to the dipper data with parametric bootstrapping. We also estimate the number of dippers using the Horvitz-Thompson population estimator. Finally, we manipulate the dipper data in a way that allows us to illustrate a simple multistate analysis (chapter 7).

Figure 9.1. Steven C. Amstrup gets acquainted with a polar bear cub (*Ursus maritimus*) on the sea ice north of Prudhoe Bay, Alaska, prior to application of ear tags and lip tattoos. (Photo by Geoff York)

After treatment of the dipper data, we work through a more complicated open-population model (chapter 5) that provides estimates of the number of polar bears in the Southern Beaufort Sea. Finally, we illustrate a dead-recovery analysis in program MARK using data from a study of mallard ducks where tag recoveries were made during 13 hunting seasons (chapter 6).

Although covering only a portion of the depth and breadth of capture–recapture studies, these examples should firm up the reader's understanding of the what, why, and how of capture–recapture. Hopefully, readers will see similarities between one of these examples and their own data and therefore feel more comfortable about carrying out an analysis.

9.2 Open-population Analyses of Data on the European Dipper

Data from the European dipper study (Marzolin 1988) have been analyzed several times in the capture–recapture literature (e.g., Lebreton et al. 1992; Cooch et al. 1996; Cooch and White 2001; Arnason and Schwarz 1998). The European dipper is a passerine bird that feeds primarily on aquatic insects in montane riparian zones. During seven annual occasions between 1981 and 1987, 294 breeding adult dippers of known sex were captured and marked in an approximately 2200-km^2 area of eastern France. Some male and female model parameters are expected to be the same due to the fact that most birds were captured as breeding pairs (Lebreton et al. 1992). Environmental factors such as the flooding or freezing of streams are also known to impact dipper population dynamics, and a major flood did occur in the study area during the 1983 breeding season (Lebreton et al. 1992).

The individual capture histories of European dippers can be presented as

$$1111110_{\sqcup}1_{\sqcup}0$$

$$1111100_{\sqcup}0_{\sqcup}1$$

$$1111000_{\sqcup}1_{\sqcup}0$$

where the seven consecutive 1's and 0's indicate whether the bird was captured (1) or not (0) at each sampling occasion, the two "$_{\sqcup}$" represent spaces and were inserted to emphasize their presence in the MARK input file format, a 1 in the next column to the right indicates that the bird is a member of group 1 (males), and a 1 in the last column to the right indicates that the bird is a member of group 2 (females). For example, the capture history $1111000_{\sqcup}1_{\sqcup}0$ represents a male dipper that was captured on the first four occasions and not captured on the final three occasions. This is the format of the program MARK input file "DIPPER.INP." This data file is available at the MARK web site: http://www.phidot.org/software/mark/docs/book/. Table 9.1 presents a condensed version of the entire dipper data set in which the observed capture histories are followed by the number of birds in each group (male or female) that had that capture history.

The European dipper data can also be summarized in the form of m-arrays (table 9.2) that list the number of first recaptures from each release cohort. Here R_j denotes the total number of birds captured and released on occasion j (i.e., number in release cohort j), and $m_{jj'}$ denotes the number of birds released on occasion j that were first recaptured on subsequent occasion j'. For example, $m_{12} = 6$ in the male m-array contained in table 9.2 indicates that, of the 12 male birds released on occasion $j = 1$, six were first recaptured on occasion $j' = 2$. If some of these six were

TABLE 9.1
Capture histories for 294 European dippers captured between 1981 and 1987
in eastern France

Capture history	Number of individuals with capture history	
	Males	Females
0000001	17	22
0000010	11	12
0000011	12	11
0000100	9	7
0000110	3	6
0000111	10	6
0001000	6	10
0001001	1	1
0001011	0	1
0001100	6	5
0001110	3	4
0001111	6	2
0010000	11	18
0010110	1	0
0011000	8	4
0011100	4	2
0011110	1	1
0011111	0	2
0100000	11	18
0110000	7	4
0110110	0	1
0111000	1	1
0111100	1	2
0111110	0	1
0111111	0	2
1000000	5	4
1010000	1	1
1100000	4	2
1101110	0	1
1111000	1	1
1111100	0	1
1111110	1	0

Source. Data from Marzolin (1988).

TABLE 9.2
Observed *m*-arrays for the European dipper data in table 9.1

Sex	*j*	*Number released* (R_j)	$m_{jj'}$					
			$j' = 2$	3	4	5	6	7
Males	1	12	6	1	0	0	0	0
	2	26		11	0	0	0	0
	3	37			17	1	0	0
	4	39				22	0	1
	5	45					25	0
	6	48						28
	7	46						
Females	1	10	5	1	0	0	0	0
	2	34		13	1	0	0	0
	3	41			17	1	0	0
	4	41				23	1	1
	5	43					26	0
	6	50						24
	7	47						
Both	1	22	11	2	0	0	0	0
	2	60		24	1	0	0	0
	3	78			34	2	0	0
	4	80				45	1	2
	5	88					51	0
	6	98						52
	7	93						

seen again on occasion 3, they are included in the count m_{23}, because they were captured and released on occasion 2 and first recaptured on occasion 3. As another example, $m_{47} = 1$ in the male *m*-array indicates that, of the 39 male birds captured and released on occasion 4, one was first recaptured on occasion 7. Because no dippers were lost on capture, the number of individuals captured and released alive at each occasion (R_j in table 9.2) is identical to the number of individuals captured at each occasion (n_j).

Recall from chapter 3 that the distinction between the CJS and the JS models is that the CJS model is conditioned on first capture. Therefore, the parameters estimated by CJS models apply only to the subset of animals that were captured at least once. The main parameters of the CJS model are ϕ_j and p_j, where ϕ_j is the apparent survival probability of animals alive just after capture occasion j to capture occasion $j + 1$, and p_j is the probability of capture on occasion j. The ϕ_j parameters of the CJS model are referred to as apparent survival probabilities because they do not differentiate between death and emigration.

The JS and Manly-Parr models, unlike the CJS model, assume that all animals in the population have the same probability of being captured as those that actually were captured. The JS and Manly-Parr models therefore permit the estimation of parameters related to the entire population, such as the abundance at occasion j (N_j) and the number of animals that entered the population (e.g., through birth or immigration) and are available for capture between occasions j and $j + 1$ (B_j). The following two sections illustrate how to fit closed-form versions of the standard JS and Manly-Parr models to the dipper data. Because these models are straightforward and can be computed by hand, they provide insight into capture–recapture modeling that may be difficult to obtain by using the specialized software that is available for more complicated capture–recapture analyses (e.g., program MARK [White and Burnham 1999], program JOLLY [Pollock et al. 1990]).

Jolly-Seber Estimates

To derive JS estimators of survival probabilities, capture probabilities, and abundance (chapter 3), it must be assumed that marked and unmarked dippers were randomly sampled from the population with a constant probability of capture for all individuals at each capture occasion. The JS estimates are functions of the following statistics: the number of animals released alive at occasion j (R_j), the number of animals seen prior to occasion j, not seen at j, and seen again after j (z_j), the number of animals released at occasion j and later recaptured (r_j), and the total number of marked animals recaptured at occasion j (m_j). All of these statistics can be derived from the observed m-arrays. First, r_j can be computed as the total of row j. Next, m_j can be computed as the total of column j. Finally, z_j is the sum of the cells in the rectangular upper submatrix of cell ($j - 1, j + 1$), i.e., the sum of contents in all cells with a row index less than or equal to $j - 1$ and a column index greater than or equal to $j + 1$. For example, z_2 is the sum of the upper submatrix of cell (1, 3): $z_2 = m_{13} + m_{14} + m_{15} + m_{16} + m_{17}$. The m-array summed over both sexes in table 9.2 yields the following statistics: $r_1 = 11 + 2 = 13$, $r_2 = 24 + 1 = 25$, $r_3 = 34 + 2 = 36$, $r_4 = 45 + 1 + 2 = 48$, $r_5 = 51$, $r_6 = 52$, $m_1 = 0$, $m_2 = 11$, $m_3 = 26$, $m_4 = 35$, $m_5 = 47$, $m_6 = 52$, $m_7 = 54$, $z_2 = 2$, $z_3 = 1$, $z_4 = 2$, $z_5 = 3$, and $z_6 = 2$.

The parameters of biological interest under the JS model (e.g., ϕ_j, p_j, and N_j) can be explicitly estimated by plugging the statistics that were calculated above into equations 3.1, 3.2, and 3.3. Standard errors are calculated by substituting estimated and sample quantities into equations 3.6 and 3.7. For example, $\hat{N}_j$ replaces N_j, $\hat{M}_j$ replaces M_j, n_j replaces $E(n_j)$, m_j replaces $E(m_j)$, and r_j replaces $E(r_j)$ in these equations. Table 9.3

TABLE 9.3

Jolly-Seber estimates of survival and capture probabilities and abundance for the European dipper example

Occasion i	Number released and number captured R_i and n_i	Number marked m_i	Number of marked seen after j r_j	Number seen before j, not seen at j, and seen after j z_j	Estimated number of marks $\hat{M}_i$	Survival estimate $\hat{\phi}_i$	Standard error $SE(\hat{\phi}_i)$	Abundance estimate $\hat{N}_i$	Standard error $SE(\hat{N}_i)$
1	22	0	13	—	0.00	0.7182	0.1555	—	—
2	60	11	25	2	15.80	0.4347	0.0688	86.18	19.58
3	78	26	36	1	28.17	0.4782	0.0597	84.50	6.12
4	80	35	48	2	38.33	0.6261	0.0593	87.62	4.78
5	88	47	51	3	52.18	0.5985	0.0561	97.69	4.83
6	98	52	52	2	55.77	—	—	105.10	4.36
7	93	54	—	—	—	—	—	—	—

shows the JS statistics and parameter estimates for the pooled dipper data set (males and females analyzed together). Note that this standard JS model allows survival and capture probabilities to vary among occasions, but that because males and females are pooled, differences between sexes are not allowed. Using the nomenclature introduced in chapter 5, this could be described as model (ϕ_t, p_t).

Manly-Parr Estimates

The Manly-Parr model is similar to the JS model in that it estimates parameters related to the entire population (e.g., ϕ_j, p_j, and N_j), and therefore requires the assumption that all individuals have an equal probability of capture on each occasion. The theoretical basis of the Manly-Parr model lies in its intuitive estimator of capture probability (p_j). This estimator requires the computation of two statistics: the number of animals known to be alive at each occasion (C_j), and the number of animals known to be alive that were actually captured at each occasion (c_j). The estimated probability of capture on occasion j (p_j) is simply the ratio (c_j/C_j). The statistics C_j and c_j cannot be derived from m-arrays, and are most conveniently computed from raw capture histories such as those shown in table 9.1.

We illustrate computation of the Manly-Parr estimated capture probability by considering occasion $j = 3$. The number of animals known to be alive at occasion 3 (C_3) is defined as the number of animals that were caught both before and after occasion 3. For the pooled dipper data set, these animals are represented by the capture histories: 0110110, 0111000, 0111100, 0111110, 0111111, 1101110, 1111000, 1111100, and 1111110. C_3 is calculated by summing the frequencies of these histories: $C_3 = 1 + 2 + 3 + 1 + 2 + 1 + 2 + 1 + 1 = 14$. Of the 14 dippers known to be alive at occasion 3, all but one, represented by the capture history 1101110, were actually captured at occasion 3. Consequently, $c_3 = 13$. The Manly-Parr estimate of capture probability is therefore $\hat{p}_3 = c_3/C_3 = 13/14 = 0.9286$. The Manly-Parr estimate of abundance at occasion 3 is calculated as $\hat{N}_3 = n_3/\hat{p}_3 = 78/0.9286 = 84.0$. Table 9.4 shows the Manly-Parr statistics and parameter estimates for the pooled dipper data set. As with the JS model, no abundance estimates are available for the first and last occasions.

The estimated standard errors that appear in table 9.4 were computed using the equation

$$SE(\tilde{N}_j) = \sqrt{\frac{\tilde{N}_j(C_j - c_j)(n_j - c_j)}{c_j^2}}$$

TABLE 9.4

Manly-Parr estimates of number of animals known to be alive, capture probability, and abundance from the European dipper data

Occasion	Number captured n_j	Number known alive C_j	Known alive that were captured c_j	Estimated capture probability $\hat{p}_j$	Estimated population size $\hat{N}_j$	Standard error of population size $SE(\hat{N}_j)$
2	60	7	5	0.714	84.00	19.22
3	78	14	13	0.929	84.00	5.68
4	80	21	19	0.905	88.42	5.45
5	88	29	26	0.897	98.15	5.20
6	98	31	29	0.935	104.76	4.15

(Seber 1982, p. 233). The JS and Manly-Parr estimates of abundance and their standard errors are remarkably similar, with an average difference in abundance equal to 0.35 animals.

Cormack-Jolly-Seber Models

The CJS model provides a flexible framework for conditional open-population modeling. Program MARK (White and Burnham 1999) is the most commonly used specialized software for CJS analyses, and provides a user-friendly environment in which CJS (and many other) models can be built that incorporate explanatory covariates. Covariates can be thought of as any additional data (e.g., the size individual animals, yearly weather patterns, etc.) that may help to explain the observed capture histories. In the case of the European dipper, for example, the sex of animals captured and the flood of 1983 are covariates that may help explain variation in survival and capture probabilities, and hence in individual capture histories.

Here we first show how to perform a basic CJS analysis of the European dipper data using the Parameter Index Matrix (PIM) approach of program MARK. The PIMs are the easiest way to build models in program MARK, and allow the specification of simple models in which separate parameters are estimated for different groups (e.g., males and females in the dipper example). We then illustrate how to model the dipper data using the design matrix approach in MARK. Parameter constraints, confounded parameters, and reduced models are the main topics of this section. Our goal is not to present a comprehensive analysis of the dipper data (see Lebreton et al. 1992; Cooch and White 2001), but rather to use real data to illustrate the mechanics of running models and interpreting output in program MARK. We complete the CJS treatment of the dipper data by introducing a general regression approach to modeling capture–recapture data. This general regression approach facilitates dealing with complex covariate structures, and is a flexible and intuitive alternative to the design matrix approach used in program MARK.

USING PROGRAM MARK

To begin analyses, start MARK and create a new project by selecting *New* from the *File* drop-down menu. Name the project and use the *Click to Select File* button to locate the DIPPER.INP input file and confirm that it is formatted correctly using the *View File* button. Designate a CJS analysis by selecting *Recaptures Only* from the radio-button list to the left. Specify 7 sampling occasions and 2 attribute groups. Click the *Enter Group Labels* button and enter Males for Group 1 and Females for Group 2. Click *OK* in the *Enter Specifications for MARK Analysis* window to read the data into MARK and begin the analysis.

Many mark–recapture analyses begin with the construction of a global model, which is a general model that has a relatively large number of parameters and is thought to be consistent with known biology and causal relationships. A global model for the dipper data is one that allows capture and survival probabilities to vary by year (t) and by sex (s). Using the nomenclature of Lebreton et al. (1992), this model is denoted (ϕ_{s*t}, p_{s*t}).

Model (ϕ_{s*t}, p_{s*t}) can be fitted using the default PIMs in MARK. PIMs are triangular matrices used to specify the number of parameters that will be estimated in a model. After reading the dipper dataset into MARK, the PIM for the apparent survival of males is automatically displayed. The default PIMs created by MARK for the other parameters can be opened by selecting *Open Parameter Index Matrix* from the *PIM* drop-down menu on the main toolbar, clicking the *Select All* button on the right, and clicking *OK*. There are a total of 4 PIMs for a two-group CJS model, and each PIM represents the structure of one parameter (ϕ_i or p_i) for one group (males or females).

Table 9.5 is a representation of the parameter indexing that MARK uses for the CJS model (ϕ_{s*t}, p_{s*t}). There are six survival parameters for each sex, corresponding to the six intervals between the seven sampling occasions of the study. Similarly there are six capture parameters for each sex, corresponding to the six possible recapture occasions. Because the CJS model conditions on first capture, it is not possible to estimate a capture probability for the first sampling occasion. MARK indexes the survival parameters first as 1–6 for males and 7–12 for females. The capture probabilities are indexed as 13–18 for males, and 19–24 for females.

TABLE 9.5
Parameter structure for European dipper model (ϕ_{s*t}, p_{s*t})

	Occasion	1	2	3	4	5	6	7
Males	Parameter	ϕ_1	ϕ_2	ϕ_3	ϕ_4	ϕ_5	ϕ_6	
	Parameter index	1	2	3	4	5	6	
Females	Parameter	ϕ_1	ϕ_2	ϕ_3	ϕ_4	ϕ_5	ϕ_6	
	Parameter index	7	8	9	10	11	12	
Males	Parameter		p_2	p_3	p_4	p_5	p_6	p_7
	Parameter index		13	14	15	16	17	18
Females	Parameter		p_2	p_3	p_4	p_5	p_6	p_7
	Parameter index		19	20	21	22	23	24

Note. There are six survival parameters for each sex, corresponding to the six intervals between the seven sampling occasions of the study, and six capture parameters for each sex, corresponding to the six possible recapture occasions. Note that it is not possible to estimate capture probability for the first sampling occasion.

TABLE 9.6
MARK parameter index matrices (PIMs) for the European dipper model (ϕ_{s*t}, p_{s*t})

Apparent survival parameter (phi) of male live recaptures (CJS)						Recapture parameter (p) of male live recaptures (CJS)					
1	2	3	4	5	6	13	14	15	16	17	18
	2	3	4	5	6		14	15	16	17	18
		3	4	5	6			15	16	17	18
			4	5	6				16	17	18
				5	6					17	18
					6						18

Apparent survival parameter (phi) of female live recaptures (CJS)						Recapture parameter (p) of female live recaptures (CJS)					
7	8	9	10	11	12	19	20	21	22	23	24
	8	9	10	11	12		20	21	22	23	24
		9	10	11	12			21	22	23	24
			10	11	12				22	23	24
				11	12					23	24
					12						24

Note. The columns in the survival PIMs represent survival rates between occasions (1–6), and the columns in the capture PIMs represent recapture probabilities at each occasion (2–7).

Once this parameterization is understood, the four PIMs shown in table 9.6 should make sense. Column k in the survival PIMs represents the index used by MARK for the survival probability over the interval k. For example, the 1 in the first column of the male survival PIM is the parameter index used for ϕ_1, which is the apparent survival probability of male dippers from occasion 1 to occasion 2. Similarly, column k in the capture PIMs represents the index used by program MARK for the capture probability at occasion $k + 1$. For example, the 21 in the third column of the female capture PIM is the parameter index for p_4, the capture probability for female dippers at occasion 4. MARK sets things up this way because the survival probability can be estimated for the first interval (ϕ_1), but the first capture probability that can be estimated is for the second occasion (p_2). Studying the PIMs and the parameter structure in table 9.5 should clarify this distinction. Note that in this example, the parameter indices are constant in any given column. This means that neither survival nor capture probabilities are allowed to vary by age of the animals or the cohort into which they were born.

It is important to realize that the numbers in the PIMs are simply indices that MARK uses to keep track of the parameters that are estimated by the model. For example, if you believe that survival does not change with time and want MARK to estimate a single survival parameter for the entire study, the same index value should be entered in every column of the survival PIM. Similarly, if you want MARK to estimate the same capture probability for each occasion of a study, the same index should be entered in each column of the capture PIM. Variation in survival or capture probabilities among cohorts (e.g., if a particular year-class had consistently poor survival because of poor nutrition in its natal year) can be modeled by using different row indices. If animals were assumed to have different survival rates following their first capture, the lower diagonals of the survival PIMs would have indices that differ from those in the remainder of the columns. See Cooch and White (2001) for a more detailed explanation of the use of PIMs.

Because no modification of the default PIMs is required in the current example, model (ϕ_{s*t}, p_{s*t}) can now be run. Select *Current Model* from the *Run* drop-down menu at the top of the screen. This opens the *Setup Numerical Estimation Run* window. Specify the name of the current model as Phi(s*t)p(s*t), and then click *OK to Run* without changing any of the default selections. The *Results Browser* window now displays a single row with the name of the current model and six other columns of information that are the basis for comparison among models. These will be considered later. Results for the current model are accessed through the buttons at the top of the *Results Browser* window. To open a Notepad window containing the real parameter estimates (the estimates of the capture and survival probabilities) and associated standard errors, click the fourth button from the left, *View estimates of real parameters in Notepad Window*. The results should be identical to those shown in table 9.7.

Two aspects of the results shown in table 9.7 require additional explanation. First, the estimate of p_3 for males is stated to be 1.000 with a standard error of 0.000. Program MARK often returns parameter estimates of 0 or 1 with corresponding standard errors that are very large or very small. In general, the user must assess such results with care. Parameter estimates that are either 0 or 1 could be caused by some aspects of the data (e.g., sparseness or structural features), which prevent an otherwise estimable parameter from being calculated. Alternatively, the parameter in question could be inherently confounded with another parameter (see below), in which case, it cannot be estimated. In the case of the male capture probability, p_3, the estimated value of 1.000 simply means that, according to the data, all of the males that could have been captured at occasion 3 were captured. This subject will be visited later, when alternative methods for specifying model (ϕ_{s*t}, p_{s*t}) through the design matrix are discussed.

The second aspect of the results in table 9.7 requiring clarification is that MARK also returns standard error estimates that are either extremely large or 0 for parameters that cannot be estimated. This is apparent in the estimates of ϕ_6 and p_7 for both males and females. Under the fully

TABLE 9.7
Survival rate and capture probability estimates from model (ϕ_{s*t}, p_{s*t}) for the European dipper data

Parameter	Occasion[a]	Estimate	Standard error	95% confidence interval[b] Lower	Upper
		Survival			
Male	1 (1981–82)	0.6970	0.2050	0.2555	0.9391
	2 (1982–83)	0.4231	0.0969	0.2520	0.6149
	3 (1983–84)	0.5053	0.0875	0.3396	0.6698
	4 (1984–85)	0.6094	0.0838	0.4389	0.7568
	5 (1985–86)	0.5708	0.0777	0.4167	0.7123
	6 (1986–87)	0.7638	0.0000	0.7638	0.7638
Female	1 (1981–82)	0.7429	0.2372	0.2021	0.9705
	2 (1982–83)	0.4468	0.0983	0.2704	0.6378
	3 (1983–84)	0.4538	0.0815	0.3036	0.6129
	4 (1984–85)	0.6404	0.0832	0.4672	0.7834
	5 (1985–86)	0.6280	0.0811	0.4610	0.7692
	6 (1986–87)	0.6928	0.0000	0.6928	0.6928
		Capture Probability			
Male	2 (1982)	0.7174	0.2239	0.2257	0.9567
	3 (1983)	1.0000	0.0000	1.0000	1.0000
	4 (1984)	0.9093	0.0856	0.5674	0.9871
	5 (1985)	0.9274	0.0693	0.6291	0.9897
	6 (1986)	0.9358	0.0617	0.6608	0.9909
	7 (1987)	0.7638	0.0000	0.7638	0.7638
Female	2 (1982)	0.6731	0.2452	0.1881	0.9482
	3 (1983)	0.8601	0.1264	0.4397	0.9796
	4 (1984)	0.9165	0.0792	0.5908	0.9882
	5 (1985)	0.8789	0.0794	0.6270	0.9691
	6 (1986)	0.9284	0.0684	0.6330	0.9898
	7 (1987)	0.6928	0.0000	0.6928	0.6928

[a] Survival estimates apply to the interval between occasion j and $j + 1$.

[b] Confidence intervals are calculated by back transforming the logit-link function. See polar bear example.

time-dependent CJS model, the terminal survival and capture probabilities are confounded and therefore cannot be individually estimated. Program MARK apparently handles this by returning the product of the terminal survival and capture probabilities. For example, the value of 0.7638 for male dipper parameters ϕ_6 and p_7 actually represents the product (ϕ_6*p_7). This product cannot be separated into its component survival and capture probabilities.

An additional caveat is required concerning the standard errors that program MARK estimates for the confounded terminal survival and capture probabilities. The magnitudes of these standard errors are not biologically or statistically meaningful. They also can vary depending on the model parameterization or even the format in which the data are read into program MARK. For example, the dipper input file containing individual capture histories (DIPPER.INP) was used to generate results in table 9.7. The standard errors for all of the confounded parameters in this table have a value of 0.000. If the exact same model is specified through the PIMs, but the condensed version of the dipper capture histories (table 9.1) is used, the standard errors for the confounded parameters are very different. For males, the standard errors of ϕ_6 and p_7 are 162.851, for females the standard errors of ϕ_6 and p_7 are 70.181. The point here is that the user must be aware of which parameters are confounded *before* running a specific model. The standard errors estimated by MARK for these parameters should be interpreted as flags that help to point out identifiability problems, not as biologically meaningful values. Chapter 4 of Cooch and White (2001) discusses this issue in more detail.

THE DESIGN MATRIX APPROACH IN PROGRAM MARK

Now we consider two methods for analyzing the dipper data via the design matrix in MARK. The first method builds on the CJS model previously specified through the PIMs, following the analysis presented in chapter 7 of Cooch and White (2001) but with minor modifications to improve the way that parameters are estimated. The second method uses a reformatted version of the dipper data and builds a design matrix based on individual covariates. We present this second method for two reasons. First, basing the analysis on an individual covariate (in this case the covariate sex_i to code for the sex of each animal i) presents a useful technique that is easily generalized to other individual covariates. Second, basing the analysis on an individual covariate provides a smooth transition to the general regression approach to capture–recapture, which is described later in this chapter.

The default design matrix for a given model is determined by the parameter structure that is specified in the PIMs. This design matrix can then be modified to create more complex parameter structures than are possible through the PIMs alone, such as models incorporating multiple covariates or additive effects. For a comprehensive treatment of the design matrix, the reader is referred to chapter 7 of Cooch and White (2001).

Continuing the dipper analysis, return to the same MARK project used in the PIM example, and open the default design matrix for model (ϕ_{s*t}, p_{s*t}) by selecting *Full* from the *Design* pull-down menu at the top of the screen. This design matrix is reproduced here as table 9.8A.

TABLE 9.8A

The MARK default design matrix for the European dipper model (ϕ_{s*t}, p_{s*t})

B1	B2	B3	B4	B5	B6	B7	B8	B9	B10	B11	B12	Param	B13	B14	B15	B16	B17	B18	B19	B20	B21	B22	B23	B24
1	1	1	0	0	0	0	1	0	0	0	0	1:Phi	0	0	0	0	0	0	0	0	0	0	0	0
1	1	0	1	0	0	0	0	1	0	0	0	2:Phi	0	0	0	0	0	0	0	0	0	0	0	0
1	1	0	0	1	0	0	0	0	1	0	0	3:Phi	0	0	0	0	0	0	0	0	0	0	0	0
1	1	0	0	0	1	0	0	0	0	1	0	4:Phi	0	0	0	0	0	0	0	0	0	0	0	0
1	1	0	0	0	0	1	0	0	0	0	1	5:Phi	0	0	0	0	0	0	0	0	0	0	0	0
1	1	0	0	0	0	0	0	0	0	0	0	6:Phi	0	0	0	0	0	0	0	0	0	0	0	0
1	0	1	0	0	0	0	0	0	0	0	0	7:Phi	0	0	0	0	0	0	0	0	0	0	0	0
1	0	0	1	0	0	0	0	0	0	0	0	8:Phi	0	0	0	0	0	0	0	0	0	0	0	0
1	0	0	0	1	0	0	0	0	0	0	0	9:Phi	0	0	0	0	0	0	0	0	0	0	0	0
1	0	0	0	0	1	0	0	0	0	0	0	10:Phi	0	0	0	0	0	0	0	0	0	0	0	0
1	0	0	0	0	0	1	0	0	0	0	0	11:Phi	0	0	0	0	0	0	0	0	0	0	0	0
1	0	0	0	0	0	0	0	0	0	0	0	12:Phi	0	0	0	0	0	0	0	0	0	0	0	0
0	0	0	0	0	0	0	0	0	0	0	0	13:p	1	1	1	0	0	0	0	1	0	0	0	0
0	0	0	0	0	0	0	0	0	0	0	0	14:p	1	1	0	1	0	0	0	0	1	0	0	0
0	0	0	0	0	0	0	0	0	0	0	0	15:p	1	1	0	0	1	0	0	0	0	1	0	0
0	0	0	0	0	0	0	0	0	0	0	0	16:p	1	1	0	0	0	1	0	0	0	0	1	0
0	0	0	0	0	0	0	0	0	0	0	0	17:p	1	1	0	0	0	0	1	0	0	0	0	1
0	0	0	0	0	0	0	0	0	0	0	0	18:p	1	1	0	0	0	0	0	0	0	0	0	0
0	0	0	0	0	0	0	0	0	0	0	0	19:p	1	0	1	0	0	0	0	0	0	0	0	0
0	0	0	0	0	0	0	0	0	0	0	0	20:p	1	0	0	1	0	0	0	0	0	0	0	0
0	0	0	0	0	0	0	0	0	0	0	0	21:p	1	0	0	0	1	0	0	0	0	0	0	0
0	0	0	0	0	0	0	0	0	0	0	0	22:p	1	0	0	0	0	1	0	0	0	0	0	0
0	0	0	0	0	0	0	0	0	0	0	0	23:p	1	0	0	0	0	0	1	0	0	0	0	0
0	0	0	0	0	0	0	0	0	0	0	0	24:p	1	0	0	0	0	0	0	0	0	0	0	0

Note. Param indicates real parameters (survival and capture probabilities), while B represents beta coefficients. Note that this parameterization will not provide correct estimates of standard errors of the beta coefficients.

Table 9.8A can be viewed as a direct representation of the linear equations used to model survival and capture probabilities. In other words, the design matrix expresses the CJS parameters that we are interested in (ϕ_j and p_j) as linear functions of beta coefficients. Each row k of the design matrix corresponds to the real parameter (ϕ_j or p_j) with index value k, as specified in the PIMs. For example, row 9 of the design matrix in table 9.8A corresponds to the parameter with index value of 9, which is the female survival probability ϕ_3. The columns of the design matrix correspond to the beta coefficients used to model survival and capture probability parameters. For example, row 9 of the design matrix in table 9.8A shows that the female survival parameter ϕ_3 is modeled by the beta coefficients B1 and B5. This means that the linear equation for the female parameter ϕ_3 can be written as

$$\text{logit}(\phi_3) = B1 + B5$$

where $\text{logit}(\phi_3)$ is a transformation of ϕ_3 that is discussed in more detail later, and B1 and B5 are the actual beta coefficients that program MARK will estimate and then use to back-calculate ϕ_3. Here B1 is a constant, or intercept, term on which all survival probabilities are based, and B5 is a time effect particular to survival over the third interval. As another example, the male dipper capture probability p_2 (the probability of capture at occasion 2) has an index value of 13, and therefore corresponds to row 13 of table 9.8. The linear function for the male parameter p_2 can be written as

$$\text{logit}(p_2) = B13 + B14 + B15 + B20$$

The beta coefficients on the right-hand side of this equation correspond, in order, to a constant term for capture probabilities, a general effect for males, a time effect particular to the second sampling occasion, and an effect for males that is specific to the second sampling occasion.

Rows 1–12 of the design matrix represent the equations for the 12 survival probabilities under model being considered, while rows 13–24 represent the equations for the 12 capture probabilities. In effect, the design matrix in table 9.8A represents the same model as was defined earlier using the PIMs (table 9.6). The only difference is that the design matrix presents model (ϕ_{s*t}, p_{s*t}) in a regression context, and provides the flexibility inherent in modeling each parameter as a linear function of some combination of products of beta coefficients and covariate values.

For those readers familiar with regression problems, it is instructive to point out that the design matrix in table 9.8A can be summarized by two general regression models. The probability of survival for every animal i between occasion j and $j + 1$ (ϕ_{ij}) can be written as

$$\ln\left(\frac{\phi_{ij}}{1 - \phi_{ij}}\right) = B_1 + B_2(sex_i) + B_3 t_{i1} + \cdots + B_7 t_{i5} \tag{9.1}$$

$$+ B_8(t_{i1} sex_i) + \cdots + B_{12}(t_{i5} sex_i)$$

where ln represents the natural logarithm, the B's are beta coefficients, t_{ia} is a dummy variable coding for occasion such that $t_{ia} = 1$ if $j = a$ and $t_{ia} = 0$ if $j \neq a$, and sex_i is an individual covariate coding for sex such that $sex_i = 1$ if animal i is a male and 0 if animal i is a female. The general regression model for capture probability has the same form but involves different beta coefficients:

$$\ln\left(\frac{p_{ij}}{1 - p_{ij}}\right) = B_{13} + B_{14}(sex_i) + B_{15} t_{i2} + \cdots + B_{19} t_{i6} \tag{9.2}$$

$$+ B_{20}(t_{i2} sex_i) + \cdots + B_{24}(t_{i6} sex_i)$$

The function $\ln[x/(1 - x)]$ that appears on the left-hand side of the above equations is called the logit function. This is one of several link functions that are used to model probabilities as linear functions of covariates. Among other things, the logit function ensures that the estimated probabilities are constrained to values between 0 and 1. McCullagh and Nelder (1989) give a general description of these functions and their uses.

Unfortunately, the issue of confounded terminal parameters discussed earlier causes a problem with the default parameterization of the design matrix in MARK (table 9.8A). Consider the terminal survival parameter for females (ϕ_6), which corresponds to row 12 of the design matrix and is a linear function of the single beta coefficient B1. B1 also serves as the constant, or intercept, beta coefficient in the linear equations for all of the other survival parameters. Similarly, the terminal capture parameter for females (p_7) is a linear function of the beta coefficient that serves as the constant for the capture probability equations, B13. The problem, alluded to earlier, is that under the fully time-dependent CJS model the beta coefficients B1 and B13 are confounded, and only their product can be estimated. Because the MARK default design matrix establishes B1 and B13 as the constant terms in all of the linear equations, the lack of identifiability in those terminal parameters affects the estimation of all of the other beta coefficients. The result is that estimates for B1 to B24 appear reasonable, but have ridiculously large standard errors. This problem can be overcome by changing the parameterization of the default design matrix in table 9.8A so that the beta coefficients that serve as the constant terms can be estimated. The modified design matrix in table 9.8B specifies the same model (ϕ_{s*t}, p_{s*t}), except that with this parameterization, the constant term for survival probabilities (B1) corresponds to the female survival parameter ϕ_1 (row 7 of the design matrix), and we can individually estimate ϕ_1. Similarly, the constant term for capture probabilities (B13) now corresponds to the female capture parameter p_2 (row 19 of the design matrix), which can also be individually estimated. Consequently, the other beta coefficients (except for the terminal values) B1 to B24 as defined in table 9.8B can now be estimated and have reasonable standard errors.

With the design matrix in table 9.8B open in program MARK, run the current model by selecting *Run* from the *Run* drop-down menu at the top of the screen. The results including the AIC, Deviance, etc. are displayed in the *Results Browser: Live Recaptures (CJS)* window. The estimates for the beta coefficients B1 to B24 are shown in table 9.9. Notice that under the current parameterization the beta coefficients B7, B12, B19, and B24 are confounded and cannot be individually estimated, as indicated by the very large standard errors. In addition, the beta coefficient B20 has a very large estimate, with an even larger standard error. This relates back to the fact that program MARK is trying to make p_3 for males equal to one, which is equivalent to making B20 infinite. The estimates of the real parameters (survival and capture probabilities) for the design matrix in table 9.8B are identical to the estimates from the PIM approach in table 9.7, with the exception of the confounded parameters (ϕ_6 and p_7 for both males and females).

THE INDIVIDUAL COVARIATE APPROACH IN PROGRAM MARK

We now build a design matrix for model (ϕ_{s*t}, p_{s*t}) using the individual covariate sex_i. This requires a slightly reformatted version of the dipper input file, which is available as DIPPER_S.INP at http://www.west-inc .com. The first 5 of the 294 individual capture histories in this file look like

$$1111110_{\sqcup}1_{\sqcup}1$$

$$1111100_{\sqcup}1_{\sqcup}0$$

$$1111000_{\sqcup}1_{\sqcup}1$$

$$1111000_{\sqcup}1_{\sqcup}0$$

$$1101110_{\sqcup}1_{\sqcup}0$$

TABLE 9.8B

An alternative parameterization of the design matrix for the European dipper model (ϕ_{s*t}, p_{s*t})

B1	B2	B3	B4	B5	B6	B7	B8	B9	B10	B11	B12	Param	B13	B14	B15	B16	B17	B18	B19	B20	B21	B22	B23	B24
1	0	0	0	0	0	0	0	0	0	0	0	1:Phi	0	0	0	0	0	0	0	0	0	0	0	0
1	1	0	0	0	0	0	0	0	0	0	0	2:Phi	0	0	0	0	0	0	0	0	0	0	0	0
1	0	1	0	0	0	0	0	0	0	0	0	3:Phi	0	0	0	0	0	0	0	0	0	0	0	0
1	0	0	1	0	0	0	0	0	0	0	0	4:Phi	0	0	0	0	0	0	0	0	0	0	0	0
1	0	0	0	1	0	0	0	0	0	0	0	5:Phi	0	0	0	0	0	0	0	0	0	0	0	0
1	0	0	0	0	1	0	0	0	0	0	0	6:Phi	0	0	0	0	0	0	0	0	0	0	0	0
1	0	0	0	0	0	1	0	0	0	0	0	7:Phi	0	0	0	0	0	0	0	0	0	0	0	0
1	0	0	0	0	0	0	1	0	0	0	0	8:Phi	0	0	0	0	0	0	0	0	0	0	0	0
1	0	0	0	0	0	0	0	1	0	0	0	9:Phi	0	0	0	0	0	0	0	0	0	0	0	0
1	0	0	0	0	0	0	0	0	1	0	0	10:Phi	0	0	0	0	0	0	0	0	0	0	0	0
1	0	0	0	0	0	0	0	0	0	1	0	11:Phi	0	0	0	0	0	0	0	0	0	0	0	0
1	0	0	0	0	0	0	0	0	0	0	1	12:Phi	0	0	0	0	0	0	0	0	0	0	0	0
0	0	0	0	0	0	0	0	0	0	0	0	13:p	1	0	0	0	0	0	0	0	0	0	0	0
0	0	0	0	0	0	0	0	0	0	0	0	14:p	1	1	0	0	0	0	0	0	0	0	0	0
0	0	0	0	0	0	0	0	0	0	0	0	15:p	1	0	1	0	0	0	0	0	0	0	0	0
0	0	0	0	0	0	0	0	0	0	0	0	16:p	1	0	0	1	0	0	0	0	0	0	0	0
0	0	0	0	0	0	0	0	0	0	0	0	17:p	1	0	0	0	1	0	0	0	0	0	0	0
0	0	0	0	0	0	0	0	0	0	0	0	18:p	1	0	0	0	0	1	0	0	0	0	0	0
0	0	0	0	0	0	0	0	0	0	0	0	19:p	1	0	0	0	0	0	1	0	0	0	0	0
0	0	0	0	0	0	0	0	0	0	0	0	20:p	1	0	0	0	0	0	0	1	0	0	0	0
0	0	0	0	0	0	0	0	0	0	0	0	21:p	1	0	0	0	0	0	0	0	1	0	0	0
0	0	0	0	0	0	0	0	0	0	0	0	22:p	1	0	0	0	0	0	0	0	0	1	0	0
0	0	0	0	0	0	0	0	0	0	0	0	23:p	1	0	0	0	0	0	0	0	0	0	1	0
0	0	0	0	0	0	0	0	0	0	0	0	24:p	1	0	0	0	0	0	0	0	0	0	0	1

Note. Param indicates real parameters (survival and capture probabilities) while B represents beta coefficients. This parameterization corrects the problems in table 9.8A and results in correct estimates for standard errors of beta coefficients.

TABLE 9.9

Estimates of beta coefficients for the alternative parameterization of the design matrix for the European dipper model (ϕ_{s*t}, p_{s*t}), as shown in table 9.8B

Parameter type	Name	Number	Estimate	Standard error	95% confidence interval Lower	Upper
	intercept	B1	1.0609	1.2417	−1.3729	3.4947
	group	B2	−0.2280	1.5759	−3.3168	2.8609
Survival	t_2	B3	−1.2743	1.3759	−3.9711	1.4225
	t_3	B4	−1.2461	1.2846	−3.7639	1.2716
	t_4	B5	−0.4837	1.2933	−3.0185	2.0512
	t_5	B6	−0.5370	1.2893	−3.0641	1.9900
	t_6	B7	1.9934	463.9219	−907.2934	911.2803
	$group*t_2$	B8	0.1313	1.7782	−3.3539	3.6165
	$group*t_3$	B9	0.4344	1.6475	−2.7948	3.6635
	$group*t_4$	B10	0.0955	1.6548	−3.1478	3.3389
	$group*t_5$	B11	−0.0107	1.6446	−3.2340	3.2127
	$group*t_6$	B12	−0.7227	471.8000	−925.4507	924.0054
	intercept	B13	0.7221	1.1144	−1.4620	2.9063
	group	B14	0.2095	1.5687	−2.8651	3.2841
Capture	t_3	B15	1.0937	1.5311	−1.9073	4.0947
	t_4	B16	1.6735	1.5208	−1.3073	4.6543
	t_5	B17	1.2596	1.3411	−1.3690	3.8883
	t_6	B18	1.8397	1.5167	−1.1331	4.8125
	t_7	B19	−0.7116	42.0157	−83.0623	81.6391
	$group*t_3$	B20	14.3940	1441.2956	−2810.5454	2839.3333
	$group*t_4$	B21	−0.3000	2.1467	−4.5075	3.9075
	$group*t_5$	B22	0.3565	2.0196	−3.6019	4.3150
	$group*t_6$	B23	−0.0915	2.1388	−4.2834	4.1005
	$group*t_7$	B24	0.4188	51.7167	−100.9459	101.7836

Note. Very large standard errors still occur for the inestimable coefficients. Also note the very large value for *group*$*t_3$, which corresponds with the inaccurate estimate of male capture probability on occasion 3 reported in table 9.7.

where again the "$\sqcup$" characters emphasize the spaces required in the MARK input file.

The first 7 digits on each line are the observed capture histories, the next digit (after the space) is the frequency of the history (all 1's in this example), and the last digit is the individual covariate sex_i, which is 1 for males and 0 for females. Start and name a new project in MARK, select the input file, and specify 7 occasions. Leave the number of attribute groups at the default value of 1 and specify 1 individual covariate. Name the covariate by clicking the *Enter Ind. Cov. Names* button and typing "sex". Click *OK* on the *Enter Specifications for MARK Analysis* window to ingest the data and start the analysis.

The most obvious way to set up the design matrix for model (ϕ_{s*t}, p_{s*t}) using the individual covariate sex_i is shown in table 9.10A. This parameterization is analogous to the design matrix in table 9.8A, which was the MARK default when modeling two groups. Unsurprisingly, the design matrix in table 9.10A leads to the same identifiability problems in the beta coefficients as discussed above. Once again, the solution is to use a slightly modified version of the design matrix in table 9.10A, which avoids these problems. The preferred way to set up the design matrix for model (ϕ_{s*t}, p_{s*t}) using the individual covariate sex_i is shown in table 9.10B.

To set up this design matrix, select *Reduced* from the *Design* drop-down menu at the top of the screen, specify 24 covariate columns, and click *OK*. Fill in the blank design matrix so that it looks like table 9.10B. It is important to be comfortable with the fact that the design matrix in table 9.10B specifies the same model as the design matrix in table 9.8B. The reason that the two design matrices look different is that table 9.8B split the dippers into two groups, using one row to represent the linear equation for each male parameter at sampling occasion j, and one row to represent the linear equation for each female parameter at sampling occasion j. Table 9.10B condenses table 9.8B by incorporating the individual covariate sex_j, which assumes the value of 0 or 1, depending on the sex of each individual. Hence, a single linear equation (and therefore a single row in the design matrix) is used to represent both male and female parameters at sampling occasion j. It may help to write out the equations for several parameters as specified by tables 9.8B and 9.10B.

One more step is required before running the analysis with the design matrix in table 9.10B. By default program MARK standardizes individual covariates to have a mean of 0 and a range of approximately −3 to +3. Because it does not make sense to standardize a binary sex covariate, uncheck the box *Standardize Individual Covariates* in the *Setup Numerical Estimation Run* window. Name and run the model by clicking the *OK to Run* button at the bottom right.

The design matrix of table 9.10B yields the beta coefficient estimates in table 9.11. Note that although most of the coefficient estimates match those in table 9.9, the point and interval estimates of the confounded parameters are different. At this point there are two methods of obtaining

TABLE 9.10A

The design matrix for the European dipper model (ϕ_{s*t}, p_{s*t}) using the treatment coding scheme and the individual covariate sex_i ($0 =$ females and $1 =$ males)

B1	B2	B3	B4	B5	B6	B7	B8	B9	B10	B11	B12	Param	B13	B14	B15	B16	B17	B18	B19	B20	B21	B22	B23	B24
1	sex	1	0	0	0	0	sex	0	0	0	0	1:Phi	0	0	0	0	0	0	0	0	0	0	0	0
1	sex	0	1	0	0	0	0	sex	0	0	0	2:Phi	0	0	0	0	0	0	0	0	0	0	0	0
1	sex	0	0	1	0	0	0	0	sex	0	0	3:Phi	0	0	0	0	0	0	0	0	0	0	0	0
1	sex	0	0	0	1	0	0	0	0	sex	0	4:Phi	0	0	0	0	0	0	0	0	0	0	0	0
1	sex	0	0	0	0	1	0	0	0	0	sex	5:Phi	0	0	0	0	0	0	0	0	0	0	0	0
1	sex	0	0	0	0	0	0	0	0	0	0	6:Phi	0	0	0	0	0	0	0	0	0	0	0	0
0	0	0	0	0	0	0	0	0	0	0	0	7:p	1	sex	1	0	0	0	0	sex	0	0	0	0
0	0	0	0	0	0	0	0	0	0	0	0	8:p	1	sex	0	1	0	0	0	0	sex	0	0	0
0	0	0	0	0	0	0	0	0	0	0	0	9:p	1	sex	0	0	1	0	0	0	0	sex	0	0
0	0	0	0	0	0	0	0	0	0	0	0	10:p	1	sex	0	0	0	1	0	0	0	0	sex	0
0	0	0	0	0	0	0	0	0	0	0	0	11:p	1	sex	0	0	0	0	1	0	0	0	0	sex
0	0	0	0	0	0	0	0	0	0	0	0	12:p	1	sex	0	0	0	0	0	0	0	0	0	0

Note. Param indicates real parameter designations and B represents beta coefficients. Similar to table 9.8A, this parameterization will not provide correct estimates of standard errors of the beta coefficients.

TABLE 9.10B

An alternative parameterization of the design matrix for the European dipper model (ϕ_{s*t}, p_{s*t}) using the "treatment" coding scheme and the individual covariate sex_i (0 = females and 1 = males)

B1	B2	B3	B4	B5	B6	B7	B8	B9	B10	B11	B12	Param	B13	B14	B15	B16	B17	B18	B19	B20	B21	B22	B23	B24
1	sex	0	0	0	0	0	0	0	0	0	0	1:Phi	0	0	0	0	0	0	0	0	0	0	0	0
1	sex	1	0	0	0	0	sex	0	0	0	0	2:Phi	0	0	0	0	0	0	0	0	0	0	0	0
1	sex	0	1	0	0	0	0	sex	0	0	0	3:Phi	0	0	0	0	0	0	0	0	0	0	0	0
1	sex	0	0	1	0	0	0	0	sex	0	0	4:Phi	0	0	0	0	0	0	0	0	0	0	0	0
1	sex	0	0	0	1	0	0	0	0	sex	0	5:Phi	0	0	0	0	0	0	0	0	0	0	0	0
1	sex	0	0	0	0	1	0	0	0	0	sex	6:Phi	0	0	0	0	0	0	0	0	0	0	0	0
0	0	0	0	0	0	0	0	0	0	0	0	7:p	1	sex	0	0	0	0	0	0	0	0	0	0
0	0	0	0	0	0	0	0	0	0	0	0	8:p	1	sex	1	0	0	0	0	sex	0	0	0	0
0	0	0	0	0	0	0	0	0	0	0	0	9:p	1	sex	0	1	0	0	0	0	sex	0	0	0
0	0	0	0	0	0	0	0	0	0	0	0	10:p	1	sex	0	0	1	0	0	0	0	sex	0	0
0	0	0	0	0	0	0	0	0	0	0	0	11:p	1	sex	0	0	0	1	0	0	0	0	sex	0
0	0	0	0	0	0	0	0	0	0	0	0	12:p	1	sex	0	0	0	0	1	0	0	0	0	sex

Note. Param indicates real parameter designations, B represents beta coefficients. This parameterization corrects the problems in table 9.10A and results in correct estimates for the standard errors of the beta coefficients.

TABLE 9.11

Estimates of beta coefficients for the alternative parameterization of the design matrix for the European dipper model (ϕ_{s*t}, p_{s*t}) using the "treatment" coding scheme and the individual covariate sex_i (0 = females and 1 = males), as shown in table 9.10B

Parameter type	Name	Number	Estimate	Standard error	95% confidence interval Lower	95% confidence interval Upper
	intercept	B1	1.0609	1.2416	−1.3727	3.4944
	sex	B2	−0.2280	1.5757	−3.3164	2.8604
Survival	t_2	B3	−1.2743	1.3758	−3.9708	1.4222
	t_3	B4	−1.2462	1.2844	−3.7637	1.2713
	t_4	B5	−0.4837	1.2931	−3.0182	2.0509
	t_5	B6	−0.5370	1.2892	−3.0638	1.9897
	t_6	B7	1.8391	0.0000	1.8391	1.8391
	$sex*t_2$	B8	0.1313	1.7779	−3.3535	3.6160
	$sex*t_3$	B9	0.4344	1.6473	−2.7943	3.6631
	$sex*t_4$	B10	0.0956	1.6546	−3.1473	3.3385
	$sex*t_5$	B11	−0.0107	1.6443	−3.2336	3.2122
	$sex*t_6$	B12	−0.6281	260.6370	−511.4766	510.2203
	intercept	B13	0.7222	1.1143	−1.4619	2.9062
	sex	B14	0.2094	1.5684	−2.8646	3.2834
Capture	t_3	B15	1.0937	1.5311	−1.9072	4.0946
	t_4	B16	1.6735	1.5208	−1.3072	4.6542
	t_5	B17	1.2596	1.3411	−1.3689	3.8881
	t_6	B18	1.8397	1.5167	−1.1330	4.8123
	t_7	B19	−0.6965	0.0000	−0.6965	−0.6965
	$sex*t_3$	B20	12.9434	701.1947	−1361.3982	1387.2850
	$sex*t_4$	B21	−0.2999	2.1465	−4.5070	3.9072
	$sex*t_5$	B22	0.3565	2.0193	−3.6013	4.3144
	$sex*t_6$	B23	−0.0913	2.1385	−4.2828	4.1001
	$sex*t_7$	B24	0.4233	0.0000	0.4233	0.4233

Note. The estimate associated with male capture probability at occasion 3 is still large due to small sample size. Also note that in contrast to table 9.9, standard errors of the terminal parameters are 0 rather than very large numbers.

estimates for the survival and capture probability values in which we are ultimately interested. Which of these methods is more convenient will depend on the complexity of the analysis. The first way is to calculate survival and capture probabilities by writing out the linear equations for each parameter using the beta coefficients from table 9.11. For example, the equation for the survival of females between occasions 4 and 5 is

$$\text{logit}(\hat{\phi}_4) = \ln\left(\frac{\hat{\phi}_4}{1-\hat{\phi}_4}\right) = \text{B1} + \text{B5}$$

Solving this equation by taking the inverse of the logit function gives

$$\hat{\phi}_4 = \frac{e^{\text{B1}+\text{B5}}}{1+e^{\text{B1}+\text{B5}}} = \frac{e^{1.0609-0.4837}}{1+e^{1.0609-0.4837}} = 0.6404$$

which is equal to the estimate derived by the PIM approach in table 9.7. Similarly, the estimated probability of capture for males at occasion 6 is

$$\hat{p}_6 = \frac{e^{\text{B13}+\text{B14}+\text{B18}+\text{B23}}}{1+e^{\text{B13}+\text{B14}+\text{B18}+\text{B23}}} = \frac{e^{2.6800}}{1+e^{2.6800}} = 0.9358$$

which again agrees with the estimate in table 9.7.

To calculate the standard errors of estimated survival or capture probabilities we use the following Taylor series approximation:

$$\hat{s}_{\hat{\theta}} = \sqrt{\hat{\theta}^2(1-\hat{\theta})^2 \sum_{i=1}^{p}\sum_{j=1}^{p} x_i x_j \, \text{cov}(B_i, B_j)}$$

where $\hat{\theta}$ is the estimated survival or capture probability of interest, p is the total number of coefficients estimated in the model, x_i is the value of covariate i that went into the prediction of $\hat{\theta}$, and $\text{cov}(B_i, B_j)$ is the covariance between the ith and jth estimated coefficients. The covariances among model coefficients are available from program MARK through the drop-down menu sequence: *Output | Specific Model Output | Variance-Covariance Matrices | Beta Estimates*.

The second way to obtain point and error estimates of survival and capture probabilities for models using individual covariates is to tell MARK the specific combinations of covariates that identify the real parameter estimates in which we are interested, and then rerun the model. This is done by selecting the *User-Specified Covariate Values* option at the bottom right of the *Setup Numerical Estimation Run* window and then clicking the *OK to Run* button. MARK then prompts the user to specify the individual covariate values of interest. After running the model, real parameter estimates corresponding to the specified covariate values can be accessed from the *Results Browser* window. This method can be tedious for complex models because it provides real parameter estimates for only one combination of covariates at a time. A model with three binary individual covariates, for example, has 3^2 possible combinations of individual covariates. That model would have to be run nine times if the researcher was interested in the real parameter values for all possible covariate combinations.

REDUCED MODELS USING THE DESIGN MATRIX APPROACH
IN PROGRAM MARK

The MARK analysis of the dipper data to this point has focused on different ways to set up and run the general model (ϕ_{s*t}, p_{s*t}). The next logical step in a thorough analysis of these data would be to construct a meaningful set of candidate models. Candidate models are usually less parameterized versions of the global model, each of which represents a biologically plausible scenario. Comparisons among restricted models are intended to evaluate biological hypotheses and may reveal which effects are supported by the data. The models constructed thus far suggest that the data do not support a significant interaction between sex and time effects. The fact that the confidence intervals for the beta coefficients B8–B11 and B21–B23 in Table 9.11 all contain zero indicates that their inclusion in the current model is not justified. It would probably be necessary, however, to build and compare more candidate models using a model fit statistic such as deviance or AIC before concluding that sex has no effect on the time-varying dipper parameters and before deciding on a final model.

The final CJS analysis we would like to show in MARK is use of a design matrix to build a reduced-parameter model. Other authors (e.g., Lebreton et al. 1992) have proposed that the flood experienced in the study area in 1983 affected dipper survival probabilities in the periods 1982–83 and 1983–84. A large variety of reduced-parameter models can be built to investigate this type of hypothesis. A reasonable model for the purpose of illustration is $(\phi_{flood}, p_.)$. This three-parameter model does not include sex effects, but allows survival probability to assume one value in the flood years and another value in the nonflood years. It also constrains the capture probability to be constant across all occasions. Continuing in the same MARK project, select *Reduced* from the *Design* drop-down menu in program MARK. Specify three covariate columns, click *OK*, and modify the design matrix to look like table 9.12. By now the equations represented by this design matrix should be apparent. For example, the linear equation for those survival probabilities affected by the flood ($\hat{\phi}_2$ and $\hat{\phi}_3$) can be written

$$\text{logit}(\hat{\phi}_{flood}) = B1 + B2$$

TABLE 9.12
The MARK design matrix for the European dipper CJS reduced-parameter model (ϕ_{flood}, $p.$)

B1	B2	Param	B3
1	0	1:Phi	0
1	1	2:Phi	0
1	1	3:Phi	0
1	0	4:Phi	0
1	0	5:Phi	0
1	0	6:Phi	0
0	0	7:p	1
0	0	8:p	1
0	0	9:p	1
0	0	10:p	1
0	0	11:p	1
0	0	12:p	1

Note. This is a three-parameter model which does not delineate sex, allows apparent survival probability to assume one value in flood years and another value in nonflood years, and constrains capture probability to be constant across all occasions.

where the baseline dipper survival probability is modeled by the beta coefficient B1, and the additive effect of the flood is modeled by B2. Capture probabilities are modeled by the single beta coefficient B3.

Running the design matrix in table 9.12 results in a model with a deviance of 660.1, an increase of only 6.15 from the fully sex- and time-dependent model (ϕ_{s*t}, p_{s*t}), even though the two models differ by 21 theoretically estimable parameters. The AIC for the model (ϕ_{flood}, $p.$) is 32 lower than for model (ϕ_{s*t}, p_{s*t}). According to the AIC model fitting philosophy of Burnham and Anderson (1998), model (ϕ_{flood}, $p.$) should therefore be considered "better" than model (ϕ_{s*t}, p_{s*t}). Keep in mind that applying other model selection methods (e.g., likelihood ratio tests, or Wald t statistics to assess significance of individual coefficients) to some set of candidate models including these two models may not lead to the same conclusion. With this caveat in mind, it will be assumed here that model (ϕ_{flood}, $p.$) is the more useful representation of the dipper data.

Under model (ϕ_{flood}, p.) the estimated apparent survival probability for all dippers drops from 0.6071 in nonflood years to 0.4688 in flood years. A capture probability of 0.8998 is estimated for all occasions. Standard errors for model (ϕ_{flood}, p.) are reduced because more data are available to estimate a common value when parameters are constrained to be equal, and the apparent precision of that value is increased. We emphasize the word "apparent" here because the tightening of confidence intervals that results from constraining parameters may simply reflect the limitations imposed by the constraining assumptions. And those assumptions (e.g., no variation between sexes) may or may not be valid or reflect biological reality. Although AIC is generally considered a reasonable method of evaluating the trade-off between the apparently higher precision and increased level of bias in a reduced-parameter model, practitioners must appreciate that this trade-off is occurring and the particular way in which it is measured by AIC. While valid model results cannot be ignored, it is ultimately up to the biologist to decide whether assumptions and estimates that otherwise appear statistically appropriate are biologically sensible. Furthermore, statistics may be of little use in detecting departures from biological reality. For example, AIC cannot detect when a biologically important covariate or model form is missing from a larger set of models.

THE GENERAL REGRESSION APPROACH TO CJS MODELS

Although users can delve into the MARK design matrices and reconstruct the regression equations they represent, the process can be awkward and unintuitive. We believe it is conceptually more straightforward to start out by visualizing capture–recapture estimates in a general regression context. This approach involves assuming that the capture and survival parameters we wish to estimate are all functions of chosen covariates. To estimate the real parameters (the capture and survival probabilities), we simply estimate the coefficient values for each covariate and use them for prediction. Going directly to the regression equations can help to eliminate the black box feel of MARK and other programs in which the investigator is forced to work through intermediate steps (such as PIM or design matrices) to get to the underlying equations for the model being assumed.

When capture–recapture problems are framed this way, one matrix represents each covariate. Rather than adjusting elements in PIM and design matrices, the analyst needs to remember only that the rows in each matrix correspond to the animals (index i), and columns correspond to the capture occasions (index j). The matrix representing each covariate can then be thought of as a new page in a three-dimensional array (index k for the 3rd dimension), as illustrated in figure 9.2.

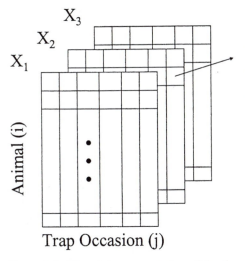

X_{ijk} = Value of k-th covariate for i-th animal on j-th trap occasion

Figure 9.2. Pictorial representation of the three-dimensional array of covariates used to parameterize capture–recapture models. Dimension 1 (rows) represents individual animals. Dimension 2 (columns) represents capture occasions. Dimension 3 (pages) represents covariates.

As an illustration, the flood model, $(\phi_{flood}, p.)$, for the dipper data fit in the previous section would need 2 matrices of covariates in the survival model, and 1 matrix of covariates in the capture model. One matrix in each model would be the intercept and is nothing but a 294×7 matrix of 1's. The second covariate in the survival model would be the flood variable matrix that has all 1's in columns corresponding to capture occasions 2 and 3. The first 5 rows of the flood covariate, corresponding to the first 5 capture histories (listed above), would be

$$flood = \begin{bmatrix} 0 & 1 & 1 & 0 & 0 & 0 & 0 \\ 0 & 1 & 1 & 0 & 0 & 0 & 0 \\ 0 & 1 & 1 & 0 & 0 & 0 & 0 \\ 0 & 1 & 1 & 0 & 0 & 0 & 0 \\ 0 & 1 & 1 & 0 & 0 & 0 & 0 \\ & & & \vdots & & & \end{bmatrix}$$

Although it was not fitted in the previous example, a sex effect could be included in model $(\phi_{flood}, p.)$ by constructing and including a sex effect matrix with element ij equal to 1 if animal i was a male, and 0 if animal i was a female. If a sex effect was desired in either the survival or capture probability model, the sex covariate matrix corresponding to the first 5 capture histories (see p. 214) would be

$$
sex =
\begin{bmatrix}
1 & 1 & 1 & 1 & 1 & 1 & 1 \\
0 & 0 & 0 & 0 & 0 & 0 & 0 \\
1 & 1 & 1 & 1 & 1 & 1 & 1 \\
0 & 0 & 0 & 0 & 0 & 0 & 0 \\
0 & 0 & 0 & 0 & 0 & 0 & 0 \\
& & & \vdots & & &
\end{bmatrix}
$$

Time (occasion) effects could also have been included using a combination of covariate matrices, named t_1 through t_7, that have all 1's in the column corresponding to the occasion they represent. The time effect matrix for occasion 3 would be

$$
t_3 =
\begin{bmatrix}
0 & 0 & 1 & 0 & 0 & 0 & 0 \\
0 & 0 & 1 & 0 & 0 & 0 & 0 \\
0 & 0 & 1 & 0 & 0 & 0 & 0 \\
0 & 0 & 1 & 0 & 0 & 0 & 0 \\
0 & 0 & 1 & 0 & 0 & 0 & 0 \\
& & & \vdots & & &
\end{bmatrix}
$$

Matrices for interaction effects between sex and time could have been fit and would have had 1's in cell ij if both the sex and time matrices had 1's in the corresponding cell. These interaction matrices can be conveniently formed using element-wise multiplication. The interaction covariate matrix for *sex* and occasion 3 (t_3) would have a 1 in column 3 for every male bird that was caught. The first 5 rows of the (t_3sex) covariate would be

$$
(t_3sex) = t_3 \times sex =
\begin{bmatrix}
0 & 0 & 1 & 0 & 0 & 0 & 0 \\
0 & 0 & 0 & 0 & 0 & 0 & 0 \\
0 & 0 & 1 & 0 & 0 & 0 & 0 \\
0 & 0 & 0 & 0 & 0 & 0 & 0 \\
0 & 0 & 0 & 0 & 0 & 0 & 0 \\
& & & \vdots & & &
\end{bmatrix}
$$

where "×" denotes element-wise multiplication.

Taken together, the set of all covariates in a capture–recapture model can be thought of as a three-dimensional array, where rows correspond to individuals, columns correspond to capture occasions, and pages correspond to covariates. In contrast to regular linear regression where each covariate value has two subscripts for the row (observation) and column (covariate), each covariate in a capture–recapture problem viewed this way has three subscripts: one for the row (the ith animal), one for the column (the jth capture occasion), and one for the page (the kth covariate). Logistic models for the survival and capture probabilities can then be generically written as

$$\ln\left(\frac{\phi_{ij}}{1-\phi_{ij}}\right) = \beta_0 + \beta_1 x_{ij1} + \beta_2 x_{ij2} + \cdots + \beta_p x_{ijp} \tag{9.3}$$

and

$$\ln\left(\frac{p_{ij}}{1-p_{ij}}\right) = \gamma_0 + \gamma_1 z_{ij1} + \gamma_2 z_{ij2} + \cdots + \gamma_q z_{ijq} \tag{9.4}$$

where x_{ija} and z_{ija} are values of the ath covariate for the ith animal during the jth capture occasion (in the case of capture probabilities) or time interval (in the case of survival probabilities), and β_a and γ_a are unknown coefficients. The capture and survival models for the $(\phi_{flood}, p.)$ model would be

$$\ln\left(\frac{\phi_{ij}}{1-\phi_{ij}}\right) = \beta_0 + \beta_i flood_{(ij)}$$

and

$$\ln\left(\frac{p_{ij}}{1-p_{ij}}\right) = \gamma_0$$

To make it easier to conceptualize the problem and to increase the utility of this method, we have, until this point, made no distinction between covariates in the capture and survival models. Rather than complicate the model specification process, we define all covariate matrices to have the same number of columns as capture occasions so that we can write them in either equation. For example, the *flood* matrix defined above with 7 columns could have been included in either the survival or capture probability model. In reality, however, only 6 columns in each covariate matrix are needed. When a covariate such as *flood* appears in the survival equation, the last column of its matrix is ignored because there are only $k - 1$ intervals between occasions when k trap occasions occur. Column j of a survival covariate matrix corresponds to the interval between capture occasion j and $j + 1$. When a covariate such as *flood* appears in the capture equation, the first column of its matrix is ignored because the first capture probability is not estimable due to conditioning of the likelihood on first captures.

Specification of very complicated models is straightforward when covariates are viewed as two-dimensional matrices because (1) PIM or design matrices do not have to be manipulated, and (2) most analysts are already familiar with the way regular regression models include complicated terms like interactions, polynomial effects, threshold effects, categorical variables, logarithmic trends, b-splines, etc. The only difference between specifying a complicated model in regular regression and specifying a complicated capture–recapture model is that the variables in the capture–recapture model are matrices instead of columns (vectors).

The general regression approach to capture–recapture modeling using three-dimensional covariate arrays offers distinct practical advantages when modeling individual covariates that can vary over time. As we have shown, the design matrix in MARK allows for specification of CJS models that incorporate time-constant individual covariates. Modeling time-dependent individual covariates in MARK, however, can be quite tedious. Time-dependent individual covariates that indicate membership in different age groups at different times, indicate known changes in reproductive status, or indicate whether an animal is wearing a radio collar can convey a significant amount of information to help explain capture histories. The only difficulty in using time-varying individual covariates is that their values must be known when the animal is not seen. Covariates that must be measured on the animal, such as litter size or herd or group size, cannot be used as time-varying individual covariates. When possible, the general regression approach to capture–recapture analyses offers an intuitive and easy way to incorporate such variables into the modeling process. Time-varying individual covariates simply have varying values in the rows and columns of their matrices.

When comparing complex models with numerous covariates using the three-dimensional matrix formulation, model differences lie entirely in the presence or absence of covariate matrices. Simply adding or subtracting covariates forms different models. Instead of adding or deleting matrices, MARK users construct a new PIM or design matrix every time a covariate is added or removed.

All that is needed to estimate the coefficients in the general capture–

recapture regression equations is a computer program that uses the covariate matrices to maximize the statistical likelihood of the data. Program MARK cannot make direct use of the covariate matrices as outlined here. The general regression approach can be carried out using the MRAWIN library of S-Plus or R routines associated with this book (available at www.west-inc.com). Behind the scenes, the MRAWIN routines utilize a Fortran program to maximize the CJS likelihood and compute estimated β and γ coefficients, standard errors, deviances, AIC, and other model fit statistics.

While these routines are free and useful for a wide range of problems, they at present have two important limitations. First, the routines are not developed to the point where they automatically compute the correct number of parameters that can be estimated, as MARK usually does. Computing the correct number of parameters is numerically difficult, and even MARK reports incorrect results with some frequency. When using AIC (or related statistics) for model selection, computation of the correct number of parameters can be a serious limitation because the correct number of parameters needs to be computed by hand. However, if AIC model selection is not being used then knowing the correct number of model parameters is not an absolute necessity. Second, to use these routines, either the S-Plus or R computer package is needed because the capture histories and all covariates need to be read into matrices using regular programming commands. Learning either the S-Plus or R language involves approximately the same amount of difficulty as learning MARK, and once learned S-Plus or R will also be beneficial to analysts for plotting and summarizing their data. The S-Plus and R languages are identical in nearly all respects, but the R package is free and downloadable from the R Project's web page (www .r-project.org).

If an analyst wishes to use these routines then all that is necessary is to construct the two-dimensional covariate matrices and invoke a single function (named *F.cr.estim*) specifying the names of the matrices containing capture histories and covariates. The routine *F.cr.estim* invokes a Fortran routine to do the actual maximization. Using the dipper data and *flood* matrix indicated above, invoking the following command fits model $(\phi_{flood}, p.)$: *fit.dipper.flood<–F.cr.estim(capture=~1, survival=~1 + flood, histories=dipper.histories)*, where *dipper.histories* is the name of the 294×7 matrix containing the dipper capture histories, and *fit.dipper.flood* is the name of the fitted capture-recapture model that is returned by the routine. After completion, typing *fit.dipper.flood* at the S-Plus or R prompt yields the following output:

Capture var	Est	SE	Survival var	Est	SE
(Intercept)	2.19438	0.33	(Intercept)	0.43508	0.13007
			flood	−0.56004	0.21651

Message = MraWin
Model df = 3
VIF = 1 on 5 df
Log likelihood = −330.0514
Deviance = 660.1028
AIC = 666.1028
QAIC = 666.1028
EBC = 660.6174

The deviance, AIC, estimated survival, and estimated capture probabilities are identical with those derived in MARK using the design matrix in table 9.12. For example, the estimated survival of dippers during non-flood years is $\exp(0.43508)/[1 + \exp(0.43508)] = 0.6071$, which matches the same result from the previous section.

INDIVIDUAL AND TIME-DEPENDENT COVARIATES IN MARK

Specifying a model with time-varying individual covariates is complicated using MARK, but it can be done. The MARK user creates a model that incorporates a time-dependent individual covariate by creating a two-dimensional matrix with the appropriate values. Then the user must think of each column of this matrix as a separate individual covariate that applies to a single occasion. In this manner, each time-dependent individual covariate for a study with j occasions can be read into MARK as j separate individual covariates. To avoid confusion and mistakes, the name of each of the j individual covariates should be explicit regarding the occasion to which it applies. Once all of the individual covariates are read into MARK, the design matrix is used to match each individual covariate with its appropriate occasion. The following example should help clarify this approach.

Consider the binary individual covariate *radio*, which indicates whether or not an animal was wearing a radio collar during some occasions over the course of a five-occasion study. The time-dependent covariate matrix *radio* has a row for each animal, and five columns. This matrix can be input into MARK by appending it to the individual capture-history matrix (such that row i of the capture-history matrix lines up with row i of the matrix *radio*) in the .INP file. After starting a MARK project, specify 5 individual covariates in the *Enter Specifications for MARK Analysis* window, and then name the covariates *radio1* through *radio5*. Table 9.13 shows how these covariates can then be entered into the design matrix to specify the simple example model (ϕ_{radio}, p_{radio}), where both survival

TABLE 9.13

The MARK design matrix for the simple example model (ϕ_{radio}, p_{radio}), where *radio* is a time-dependent individual covariate for a 5-occasion study

B1	B2	Param	B3	B4
1	radio1	1:Phi	0	0
1	radio2	2:Phi	0	0
1	radio3	3:Phi	0	0
1	radio4	4:Phi	0	0
0	0	5:p	1	radio2
0	0	6:p	1	radio3
0	0	7:p	1	radio4
0	0	8:p	1	radio5

and capture probabilities vary as a function of the *radio* covariate over the course of the study. For a short-lived study with only one covariate that varies over time and by individual, this approach in MARK can be effective. On the other hand, it would be enormously tedious in a 30-occasion study in which two or three covariates needed to be handled this way. In the general regression approach, featured in this section, a study including three variables that differed over time and by individual would require the assignment of three covariate matrices. To accomplish the same thing in MARK would require definition of 90 (3×30) individual covariates.

9.3 The Huggins Closed-population Model Applied to the European Dipper Data

In this section, we violate numerous assumptions and treat the dipper data of table 9.1 as if they came from a closed population so that we can illustrate computation of the Huggins model described in chapter 4. Nobody would consider this a proper analysis of the dipper data, but because the reader is familiar with the dipper data and the data file is already built, it makes an excellent tutorial example. The Huggins approach to modeling closed-population data is very flexible and can be used with many sets of data. We perform one analysis with MARK and another using the spreadsheet program Microsoft Excel. Because the calculations performed in the spreadsheet program are transparent, illustrating the computations that way may help the user to more completely understand the process. Using Excel, however, to complete a real analysis will rarely be feasible.

To fit the Huggins model to the dipper data in MARK, the data file used in the previous example (with one capture history per line) can be used. In MARK, a new project should be started and the *Huggins Closed Captures* model option chosen. After reading in the data specifying 7 capture occasions, and 2 groups, the default PIM matrix approach would identify full time dependence in the parameters for probability of initial capture and probability of subsequent capture. This default model would estimate a total of 26 parameters, and the reader should be able, by now, to use MARK to complete this most general model by following steps similar to the earlier dipper examples using PIMs.

Here, we want to show how to use the design matrix approach to illustrate modeling original captures as a function of one set of covariates and recaptures with another set of covariates. Sticking to the covariates discussed in earlier dipper examples, we assume that the flood differentially affects recaptures. Therefore, we modify the default design matrix in MARK to incorporate a capture model that includes a sex and time effect, and a recapture model that includes a sex and flood effect. The design matrix for this model is shown in table 9.14. Note that in this example, the sex effect we consider is a simple step effect constant across time. Hence, we are not considering possible interactions between sex and time or between sex and flood effects. Once fitted, this model estimates 141 males (SE = 1.8) and 153 females (SE = 2.3), for a total of 294 birds. Note also that the Huggins model produces only one estimate of the number of birds, not one for each capture occasion. This is because this is a model that assumes the population is not changing during the course of the study. In a real study, each covariate would be investigated for its ability to explain variation in the capture histories. Interactions between sex and time or sex and flood effects would be logical things to examine. Some variables would potentially be removed, while others would potentially be added in this exploration of the data. We will forgo the model-selection process in this example in order to focus on illustrating the underlying calculations and some fundamental ideas of maximum likelihood estimation.

The Huggins model likelihood function is relatively simple and it can be computed and maximized in a spreadsheet program like MS Excel. Although starting the spreadsheet process is a bit awkward, it is very instructive because the transparency of spreadsheet operation illuminates a fundamental idea behind all maximum likelihood techniques. All maximum likelihood techniques maximize a statistical likelihood, and the maximization process can be thought of as taking educated guesses at parameters until a certain function of the data (i.e., the likelihood) is maximized. To fit the likelihood in a spreadsheet, the spreadsheet must

TABLE 9.14
The MARK design matrix used to fit the Huggins closed-population model to the European dipper data

B1	B2	B3	B4	B5	B6	B7	B8	Param	B9	B10	B11
1	1	1	0	0	0	0	0	1:P	0	0	0
1	1	0	1	0	0	0	0	2:P	0	0	0
1	1	0	0	1	0	0	0	3:P	0	0	0
1	1	0	0	0	1	0	0	4:P	0	0	0
1	1	0	0	0	0	1	0	5:P	0	0	0
1	1	0	0	0	0	0	1	6:P	0	0	0
1	1	0	0	0	0	0	0	7:P	0	0	0
1	0	1	0	0	0	0	0	8:P	0	0	0
1	0	0	1	0	0	0	0	9:P	0	0	0
1	0	0	0	1	0	0	0	10:P	0	0	0
1	0	0	0	0	1	0	0	11:P	0	0	0
1	0	0	0	0	0	1	0	12:P	0	0	0
1	0	0	0	0	0	0	1	13:p	0	0	0
1	0	0	0	0	0	0	0	14:p	0	0	0
0	0	0	0	0	0	0	0	1:C	1	1	1
0	0	0	0	0	0	0	0	2:C	1	1	1
0	0	0	0	0	0	0	0	3:C	1	1	0
0	0	0	0	0	0	0	0	4:C	1	1	0
0	0	0	0	0	0	0	0	5:C	1	1	0
0	0	0	0	0	0	0	0	6:C	1	1	0
0	0	0	0	0	0	0	0	7:C	1	0	1
0	0	0	0	0	0	0	0	8:C	1	0	1
0	0	0	0	0	0	0	0	9:C	1	0	0
0	0	0	0	0	0	0	0	10:C	1	0	0
0	0	0	0	0	0	0	0	11:C	1	0	0
0	0	0	0	0	0	0	0	12:C	1	0	0

Note. The initial capture probability model includes sex (B2) and time (B3–B8) effects. The recapture probability model includes a separate sex (B10) and flood effect (B11). Param indicates real parameter designations and B represents beta coefficients.

have a procedure to find the values in several cells that maximize the value in another cell.

In Microsoft Excel the solver procedure can carry out the maximization. This may need to be installed for the purpose. When using any maximization routine it is best to independently verify that the true maximum was achieved. Some maximization routines have difficulty achieving the correct maximum when the function is complicated. The solver routine in MS Excel is sufficient to find the maximum of the simple Huggins likelihood illustrated here. However, analysts will probably want to use a more general purpose maximization routine on more complicated real-life problems.

We illustrate computation of the likelihood for a simple Huggins model where the initial capture probabilities change on each sampling occasion, as do recapture probabilities (i.e., there are no heterogeneity or covariates). To make this description easier, assume 4 trapping occasions and that the capture histories are contained in columns A–D of a spreadsheet, starting in row 3, as shown in figure 9.3. Also, guesses at the initial probabilities of capture are contained in cells A1, B1, C1, and D1 (more columns will be needed if more occasions are present, and the multinomial expressions described below would also be expanded accordingly if more occasions are present), and guesses at the recapture probabilities are contained in cells B2, C3, and D3. Recall that there are only three recapture probabilities that can be estimated with four sampling occasions. The following formula should be entered into cell E3 and copied down the column:

$$
\begin{aligned}
&= A3{*}LN(A\$1) + (1 - A3){*}B3{*}(LN(1 - A\$1) + LN(B\$1)) \\
&+ (1 - A3){*}(1 - B3){*}(C3){*}(LN(1 - A\$1) + LN(1 - B\$1) + LN(C\$1)) \\
&+ (1 - A3){*}(1 - B3){*}(1 - C3){*}D3{*}(LN(1 - A\$1) + LN(1 - B\$1) \\
&+ LN(1 - C\$1) + LN(D\$1))
\end{aligned}
$$

This calculates the logarithm of the likelihood of obtaining the initial captures, as explained in the *Models Incorporating Covariates* section in chapter 4. The following formula should then be entered into cell F3 and copied down the column:

$$
\begin{aligned}
&= A3{*}(B3{*}LN(B\$2) + (1 - B3){*}LN(1 - B\$2) + C3{*}LN(C\$2) \\
&+ (1 - C3){*}LN(1 - C\$2) + D3{*}LN(D\$2) \\
&+ (1 - D3){*}LN(1 - D\$2)) + (1 - A3){*}B3{*}(C3{*}LN(C\$2) \\
&+ (1 - C3){*}LN(1 - C\$2) + D3{*}LN(D\$2) \\
&+ (1 - D3){*}LN(1 - D\$2)) \\
&+ (1 - A3){*}(1 - B3){*}C3{*}(D3{*}LN(D\$2) \\
&+ (1 - D3){*}LN(1 - D\$2))
\end{aligned}
$$

	A	B	C	D	E	F	G	H
1	0.4043	0.4072	0.4579	0.4224	P part of Likelihood	C part of Likelihood	Denominator	Log (Likelihood) Contribution
2		0.4000	0.5000	0.4000				
3	0	0	0	1	−2.5151	0.0000	−0.1172	−2.3979
4	0	0	1	0	−1.8219	−0.5108	−0.1172	−2.2156
5	0	0	1	1	−1.8219	−0.9163	−0.1172	−2.6210
6	0	1	0	0	−1.4165	−1.2040	−0.1172	−2.5033
7	0	1	1	0	−1.4165	−1.2040	−0.1172	−2.5033
8	0	1	1	1	−1.4165	−1.6094	−0.1172	−2.9087
9	1	0	0	0	−0.9056	−1.7148	−0.1172	−2.5033
10	1	0	0	1	−0.9056	−2.1203	−0.1172	−2.9087
11	1	0	1	1	−0.9056	−2.1203	−0.1172	−2.9087
12	1	1	0	0	−0.9056	−2.1203	−0.1172	−2.9087
13	1	1	1	0	−0.9056	−2.1203	−0.1172	−2.9087
14								
15							Log(Likelihood) =	−29.28788

Figure 9.3. Spreadsheet used to illustrate computation of the Huggins model. Parameters are in cells A1:D2. Example capture histories are in cells A3:D13. Intermediate calculations necessary to compute the likelihood are in columns E, F, G, and H. The final log likelihood for this data and these particular parameter estimates is in cell H15.

This formula calculates the contribution of the recapture histories to the log-likelihood. Next, the following formula should be entered into cell G3 and copied down the column:

$$= LN(1 - (1 - A\$1)*(1 - B\$1)*(1 - C\$1)*(1 - D\$1))$$

This calculates the denominator of the Huggins likelihood. Lastly, = E3 + F3 − G3 should be entered into cell H3 and copied down the column. Column H contains the total contribution of each capture history to the log-likelihood. The full log-likelihood can now be calculated by summing all the rows of column H. For the particular set of 11 capture histories and parameter guesses displayed in figure 9.3, the log-likelihood is −29.29, as shown in cell H15. By changing the parameters, we wish to make this number as large as possible.

It is instructive to change the cells in the range A1:D2 (i.e., the parameters) and see how cell H15 changes. If cell H15 gets larger, the change in parameters is in the right direction because the log-likelihood is closer to its maximum. If cell H15 gets smaller, the change in parameters is in the wrong direction.

All that is left to do is continue guessing at the values in cells A1:D2 until the value in cell H15 is maximized. This is where the solver function of the spreadsheet is useful. Basically, the solver asks for a range of cells to change (i.e., A1:D2), an objective cell (i.e., H15), and whether or not to maximize or minimize. Consult the spreadsheet's help file on proper use of the solver. This model, it turns out, is over parameterized, and multiple maximum likelihood estimates exist for the same likelihood value. A solution set reported by MARK is: $P_1 = 0.4103$, $P_2 = 0.4174$, $P_3 = 0.4776$, $P_4 = 0.4571$; $C_2 = 0.4$, $C_3 = 0.5$, and $C_4 = 0.4$. The maximum likelihood value reported was -29.287. A single solution could be found using more data or different models.

A spreadsheet can easily calculate parameter point estimates as described above. Most spreadsheets, however, cannot easily compute standard errors. Whereas add-in routines for some spreadsheet programs may suffice for calculating variances, the most straightforward estimation of variances for the Huggins model are available from either program MARK or CARE-2 as described in chapter 4.

9.4 Assessing Goodness-of-Fit

How well the full CJS model fits data from each sex can be tested with goodness-of-fit (GOF) statistics. Currently, there are two common GOF methods for assessing whether the arrangement of data meets the CJS assumptions (see chapters 3 and 5). One method involves two chi-squared contingency table analyses called Test 2 and Test 3. The other method involves parametric bootstrapping. Both procedures are available in MARK, and some practical aspects of both will be described here.

Both GOF testing procedures provide two types of information: (1) a goodness-of-fit p value, and (2) an estimate of a variance inflation factor ($\hat{c}$). This $\hat{c}$ is a measure of overdispersion, or the amount of excess variance in the data due to violation of model assumptions. In other words, a model might be built on the assumption that data are arranged according to a Poisson distribution. Lack of independence, heterogeneity, or other unknown features may increase the dispersion of the data (i.e., increase the variance) beyond what would be predicted by a Poisson distribution. The estimated value for $\hat{c}$ then can be used to correct the model selection criteria for the extra variation above that predicted by the assumed model. When this is done, the analysis changes from a true maximum likelihood analysis into a quasi-likelihood analysis because the data are no longer hypothesized to come from a true likelihood. When $\hat{c}$ is incorporated into AIC calculations, we label the measure QAIC to acknowledge the quasi-likelihood nature of the statistic.

Both Test 2 and Test 3 compute the expected number of captures in cells of a contingency table based on the model of interest and compare those expected values to the actual number of captures in those cells. Test 2 examines the pattern of recaptures of animals that were captured during a particular occasion, say occasion #3. It contrasts the observed and expected number of animals captured at the first occasion following

TABLE 9.15

Chi-squared contingency table for Test 2 associated with female European dippers and occasion #3

		First recaptured at occasion 4?		
		Yes	No	Total
Seen at occasion 3?	Yes	17	1	18
	No	1	0	1
	Total	18	1	19

Note. This component table of test 2 tallies individuals captured before and at occasion #3 with those first recaptured at occasion #4 and afterward.

the one of interest (in this case #4) with the expected number of animals captured on all occasions beyond that next occasion (#5+). We count the total number of first recaptures after occasion j (#3 in this case), and then partition those into the number seen and not seen at occasion j and first recaptured at occasion $j + 1$. An example from the dipper data will help illustrate. Of the female dippers captured prior to or on occasion 3, 19 of them were later recaptured. Of those 19, 17 were seen at occasion 3 and first recaptured at occasion 4, 1 was seen at occasion 3 and not recaptured until after occasion 4, 1 was not seen at occasion 3 but was first recaptured at occasion 4, and 0 birds were not seen at occasion 3 and first recaptured after occasion 4. These counts for the female dippers recaptured after occasion 3 appear in table 9.15.

If the CJS model fits, the number of animals in each cell follows a multinomial distribution and the expected count in each cell is simply the product of the two marginal totals, divided by the grand total. In other words, if all animals have the same probability of being captured at each occasion (a simple time-dependent model), then the proportion of animals captured at occasion #3 and again at #4, should be the same as the proportion of animals captured before #3 and at #4. Similarly, the proportion of animals captured at #3 and not captured until occasion #5 or later should be the same as the number of animals captured before #3 and at #5 or later. For table 9.15, the expected count for cell (1, 1) is 18(18)/19 = 17. The expected count for cell (2, 1) is 18(1)/19 = 0.95. Using the observed and expected counts for the cells, the chi-squared statistic for the table can be computed as

$$\chi^2 = \sum_{i=1}^{2} \sum_{j=1}^{2} \frac{(o_{ij} - e_{ij})^2}{e_{ij}}$$

where o_{ij} and e_{ij} are the observed and expected counts for cell (i, j), respectively. This statistic approximately follows a chi-squared distribution (with 1 degree of freedom) under the null hypothesis that the CJS model fits. For the data in table 9.15, $\chi^2 = [(17 - 17)^2/17] + [(1 - 0.95)^2/0.95] + [(1 - 0.95)^2/0.95] + [(0 - 0.05)^2/0.05] = 0.055$.

As with all chi-square tests the approximation to the chi-square distribution may be poor if the expected cell counts are low. An accepted rule of thumb is that the chi-squared statistic is unreliable (i.e., does not follow a chi-square distribution) if any of the expected counts are <2. Because most capture–recapture datasets violate this rule of thumb for at least one or more occasions, chi-square statistics computed occasion by occasion almost always are summed across occasions and groups to compute an overall χ^2 statistic.

If the CJS model does not fit, the observed counts will be consistently far from their expected values, and the overall χ^2 statistic will be large relative to the chi-square distribution. If the area under the chi-square distribution to the right of χ^2 (i.e., the p value associated with χ^2) is less than α, we reject the null hypothesis that the CJS model adequately fits the data. The overall Test 2 chi-square statistic for female dippers is $\chi^2 = 3.2503$ (df $= 4$, $p = 0.5168$), while the overall Test 2 chi-square statistic for both males and females is $\chi^2 = 7.5342$ (df $= 6$, $p = 0.2743$). Insignificant values for both Test 2 statistics suggest no reason to believe the full CJS model does not fit the data.

Test 3 contrasts the observed and expected number of animals captured at each occasion that were seen and not seen before with those counts of animals that were seen and not seen after each occasion. Again, the idea is that animals captured at one occasion, #3, for instance, should have the same probability of being captured after occasion #3 whether or not they were captured before occasion #3. The Test 3 contingency table for the 41 female dippers captured at occasion #3 appears in table 9.16. The total number of animals in each Test 3 contingency table is the total number of animals seen at occasion j. The expected count in each cell is again the product of the two marginal totals divided by the grand total. The expected count in cell $(1, 1)$ of table 9.16 is $14(18)/41 = 6.15$. The expected count in cell $(2, 1)$ is $27(18)/41 = 11.85$. Similar to Test 2, an animal group's contingency tables and chi-square statistics for Test 3 are computed for every occasion and summed to get an overall test of fit. Test 3 for the female dippers is $\chi^2 = 2.0412$ (df $= 3$, $p = 0.5639$), while Test 3 for both male and female dippers is $\chi^2 = 10.7735$ (df $= 15$, $p = 0.7685$). As with Test 2, Test 3 does not indicate lack of fit of the full CJS model to the dipper data.

While tedious to compute by hand, Tests 2 and 3 are computed by program Release, which is a component of program MARK. To compute Tests 2 and 3 in MARK, simply read in the data file as above. Go to the

TABLE 9.16

Chi-squared contingency table for Test 3 associated with female European dippers and occasion #3

		Sean after occasion 3?		
		Yes	No	Total
Seen before occasion 3?	Yes	9	5	14
	No	9	18	27
	Total	18	23	41

Note. This component table of test 3 tallies individuals captured before and after occasion #3.

"Tests" menu and click Program RELEASE GOF. In the resulting output, all Test 3 contingency tables appear first, followed by all Test 2 contingency tables, followed by a summary of both Tests 2 and 3 together. It is this combined summary table that is usually most interesting because it has the most degrees of freedom (and is therefore most powerful), and it can be used to compute an estimate of $\hat{c}$. From the "Goodness-of-Fit Results (TEST 2 + TEST 3) by Group" table that appears in the RELEASE results applied to the dipper data, the total $\chi^2 = 18.31$ (df = 21, $p = 0.6295$). Given the high p value, results of these combined tests, like the results of Test 2 and Test 3, do not indicate lack of fit of the CJS model to the dipper data. Although both Test 2 and Test 3 rely on fairly intuitive assumptions about the distributions of captures and recaptures, they appear to have low power to actually detect a departure from the CJS model (Manly et al. 1999). Low power should be recognized when goodness-of-fit statistics are evaluated, and it suggests that development of other tests to assess model fit may be a fertile area for future research.

The best and most common estimate of $\hat{c}$ is calculated from this same table in the RELEASE output file as the overall χ^2 statistic divided by degrees of freedom. For the dipper data, $\hat{c} = 18.31/21 = 0.87$. It is customary to set $\hat{c} = 1.0$ if the actual estimate of $\hat{c} < 1$, as in this case. A $\hat{c} \leq 1$, like the GOF tests, suggests no lack of fit and therefore no adjustment to the AIC testing criteria is indicated.

Another intuitively appealing method to assess model fit is called the parametric bootstrap. The parametric bootstrap or simulation procedure uses the parameters from a selected model to create a large number of simulation data sets. Data in these simulation data sets fit the parameters of the selected model and therefore are known to meet all of the assumptions of the model (e.g., homogeneity of capture and survival probabilities).

Diagnostic statistics such as the deviance, degrees of freedom, and deviance/
degrees of freedom can then be calculated for each of the simulated data
sets. The distribution of these diagnostic values computed on the simulation data sets is then compared to the same diagnostic values calculated from the selected model when it is applied to the real data. If the
observed diagnostic statistics are abnormal when compared to those
generated by the simulated data sets, then a violation of assumptions is
suggested. For example, if the maximum deviance computed from 1000
datasets that satisfy all assumptions is 250, but the observed deviance is
400, one can reasonably infer that some assumptions of the model are
violated in the real data set. In addition, if the average deviance from
1000 data sets known to satisfy all assumptions is 200, but the observed
value is 400, then a reasonable estimate of the amount of the extra variation not accounted for by the model (the variance inflation factor) is
$\hat{c} = 400/200 = 2$.

For the dipper data, 1000 parametric simulations were performed in
MARK with the full CJS model applied to both sexes separately. The observed deviance of the fitted model was 71.5. Comparing this value to
the ranked deviances from 1000 simulations revealed that 95 of the
1000 simulated deviances were greater than or equal to 71.5. Thus, the
probability of a deviance as large or larger than the observed value is estimated to be $p = 0.095$. This probability falls short of the commonly
used significance criterion of 0.05. Statistical rules aside, a deviance this
great gives a suggestion of some violation in the assumptions of the
model. This result contrasts with the Test 2 and Test 3 results already reported, and suggests that a search for better fitting models (e.g., by testing different covariates and exploring other assumptions) is in order.

One method to estimate $\hat{c}$ with the results of parametric bootstrapping
is to divide the deviance calculated from the observed data by the mean
of the simulated deviances (Manly et al. 1999). For the dipper data, the
mean of the bootstrapped deviances is 55.66, and this ratio is $\hat{c} = 71.5/
55.66 = 1.28$. As with the significance test in the previous paragraph,
this estimate of $\hat{c}$ also indicates some lack of fit between the selected
model and the observed data. Estimated values of $\hat{c}$ can be used to correct Akaike's information criteria (AIC) for extra variation. When this is
done, the AIC becomes a quasi-AIC (QAIC) to reflect the fact that the
likelihood with the extra variation is a quasi-likelihood (Burnham and
Anderson 1988; chapter 5).

As seen above, the GOF tests and $\hat{c}$ calculations from bootstrapping
differed from those relying on Test 2 and Test 3. The latter suggested
no lack of fit and no overdispersion in the data. The bootstrapping results, on the other hand, suggested at least moderate overdispersion and
lack of fit. Because corrections for overdispersion will affect parameter

estimates provided by the model, this difference is not trivial. As pointed out previously, Manly et al. (1999) reported that Test 2 and Test 3 have low power. White (2002), on the other hand, performed simulations suggesting that bootstrapped estimates of GOF may be unreliable and that Test 2 and Test 3 are preferable. Based on the results of White (2002) and until further investigation of these techniques can be performed, we recommended that $\hat{c}$ be estimated based on the results of Test 2 and Test 3.

9.5 Horvitz-Thompson Open-population Size Estimates

As pointed out above, one difference between the JS model and the CJS model approach is that the CJS model deals only with the marked component of the population. Because it does not make inference to the unmarked portion the CJS model provides estimates of capture and survival probabilities only. Capture probabilities are, however, related to population sizes. If the investigator is willing to assume that all European dippers in the study area had positive, but not necessarily equal, capture probabilities, then the Horvitz-Thompson estimates of population size given by equation 5.5 can be computed.

Recall that a Horvitz-Thompson estimate is computed as the sum of the reciprocals of the capture probabilities for the individual animals captured on a sampling occasion. For purposes of illustrating the computations, we will apply the Horvitz-Thompson estimator using the capture probabilities shown in table 9.7. We know from the goodness-of-fit tests in the previous section and from previous analyses of the dipper data that the model used to generate the results in table 9.7 does not adequately explain the heterogeneity in the sample. To provide a simple example, however, we ignore that shortcoming, and show how to apply the Horvitz-Thompson estimator as if there were no heterogeneity issues.

Estimated capture probabilities for female dippers range from 0.67 to 0.93; note that p_7 is confounded with ϕ_6 in this model. Estimated capture probabilities for male dippers range from 0.72 to 1.0. In the model used to generate table 9.7, members of each sex have the same covariate values each sample occasion. Consequently, equation 5.5 reduces to the number of captured males and females divided by their respective capture probabilities. The estimate of population size at time $j = 2, \ldots, 6$ is therefore

$$\hat{N}_j = \sum_{i=1}^{n_j} \frac{I_{ij}}{\hat{p}_{ij}} = \frac{n_{jF}}{\hat{p}_{jF}} + \frac{n_{jM}}{\hat{p}_{jM}}$$

where n_{jF} is the number of females caught at time j, and n_{jM} is the number of males caught at time j. Both n_{jF} and n_{jM} can be found in table 9.2.

Substituting the appropriate values, the estimate of population size at time 2, when 26 male dippers and 34 female dippers were captured, is $\hat{N}_2 = 34/0.673 + 26/0.717 = 86.75$. Not surprisingly this is virtually identical to the JS population estimate shown in table 9.3. The rest of the Horvitz-Thompson estimates of population size are $\hat{N}_3 = 84.67$, $\hat{N}_4 = 87.63$, $\hat{N}_5 = 97.4$, and $\hat{N}_6 = 105.15$.

The variance of $\hat{N}_j$ can be approximated, assuming that captures of individuals are independent, using Taylor series expansions. The approximation is derived by noting that

$$var(\hat{N}_j) = var(E[\hat{N}_j | \hat{\mathbf{p}}_j]) + E[var(\hat{N}_j | \hat{\mathbf{p}}_j)] \tag{9.5}$$

(McDonald and Amstrup 2001) and approximating each term. The first term on the right-hand side of (9.5) can be decomposed as

$$var(E[\hat{N}_j | \hat{\mathbf{p}}_j]) = var\left(\sum_{i=1}^{N_j} \frac{E[I_{ij} | \hat{p}_{ij}]}{\hat{p}_{ij}}\right) = var\left(\sum_{i=1}^{N_j} \frac{p_{ij}}{\hat{p}_{ij}}\right)$$

$$= \sum_{i=1}^{N_j} \sum_{i'=1}^{N_j} p_{ij} p_{i'j} cov(1/\hat{p}_{ij}, 1/\hat{p}_{i'j})$$

Assuming $E[\hat{p}_{ij}] = p_{ij}$ (i.e., that $\hat{p}_{ij}$ is an unbiased estimate of p_{ij}), the "delta method" for approximating the covariance of a function of random variables (Seber 1982, p. 9) involves 2nd-order Taylor series expansions and yields the approximation

$$cov(1/\hat{p}_{ij}, 1/\hat{p}_{i'j}) \approx \frac{cov(\hat{p}_{ij}, \hat{p}_{i'j})}{p_{ij}^2 p_{i'j}^2}$$

so that,

$$var(E[\hat{N}_j | \hat{\mathbf{p}}_j]) \approx \sum_{i=1}^{N_j} \sum_{i'=1}^{N_j} \frac{cov(\hat{p}_{ij}, \hat{p}_{i'j})}{p_{ij} p_{i'j}} \tag{9.6}$$

If a logistic link function is used to relate the covariates to capture probabilities (i.e., $\hat{p}_{ij} = (1 + e^{-\mathbf{x}_{ij}\hat{\boldsymbol{\beta}}})^{-1}$, where $\mathbf{x}_{ij}$ is a vector of length p containing covariate values for animal i at capture occasion j and $\hat{\boldsymbol{\beta}}$ is the vector of estimated coefficients), then another application of the delta method yields

$$cov(\hat{p}_{ij}, \hat{p}_{i'j}) \approx p_{ij}(1 - p_{ij}) p_{i'j}(1 - p_{i'j}) cov(\mathbf{x}_{ij}\hat{\boldsymbol{\beta}}, \mathbf{x}_{i'j}\hat{\boldsymbol{\beta}})$$
$$= p_{ij}(1 - p_{ij}) p_{i'j}(1 - p_{i'j})(\mathbf{x}_{ij}\hat{\boldsymbol{\Sigma}}_{\boldsymbol{\beta}}\mathbf{x}'_{i'j}) \tag{9.7}$$

where $\hat{\Sigma}_\beta$ is the estimated $p \times p$ variance–covariance matrix for the coefficients in $\hat{\beta}$. Other link functions, such as arcsin, will require different approximations in equation 9.7. The other equations in this section, however, hold for all link functions.

If we assume that captures of individuals are independent of one another, and apply the 2nd-order Taylor series approximations.

$$E[1/\hat{p}_{ij}] \approx \frac{1}{p_{ij}} + \frac{var(\hat{p}_{ij})}{p_{ij}^3}$$

and

$$var(1/\hat{p}_{ij}) \approx \frac{var(\hat{p}_{ij})}{p_{ij}^4}$$

the second term of (9.5) can be decomposed as

$$E[var(\hat{N}_j | \hat{\mathbf{p}}_j)] = E\left[var\left(\sum_{i=1}^{N_j} I_{ij}/\hat{p}_{ij} | \hat{\mathbf{p}}_j \right) \right] = \sum_{i=1}^{N_j} E[(1/\hat{p}_{ij}^2)var(I_{ij} | p_{ij})]$$

$$= \sum_{i=1}^{N_j} p_{ij}(1 - p_{ij})E[1/\hat{p}_{ij}^2]$$

$$\text{(9.8)}$$

$$= \sum_{i=1}^{N_j} p_{ij}(1 - p_{ij})[var(1/\hat{p}_{ij}) + E^2(1/\hat{p}_{ij})]$$

$$\approx \sum_{i=1}^{N_j} (1 - p_{ij})\left[\frac{1}{p_{ij}} + \frac{3var(\hat{p}_{ij})}{p_{ij}^3} + \frac{[var(\hat{p}_{ij})]^2}{p_{ij}^5} \right]$$

Substituting (9.6) and (9.8) into (9.5) yields the approximate variance of $\hat{N}_j$,

$$var(\hat{N}_j) \approx \sum_{i=1}^{N_j} \frac{1 - p_{ij}}{p_{ij}}\left[1 + \frac{3var(\hat{p}_{ij})}{p_{ij}^2} + \frac{[var(\hat{p}_{ij})]^2}{p_{ij}^4} \right] + \sum_{i=1}^{N_j} \sum_{i'=1}^{N_j} \frac{cov(\hat{p}_{ij}, \hat{p}_{i'j})}{p_{ij}p_{i'j}}$$

$$\text{(9.9)}$$

To obtain an estimator of (9.9), we rely on the Horvitz-Thompson theorem and the reasoning behind its proof. When any quantity, say v_i, is obtained from a unit that is sampled at probability p_i, the Horvitz-Thompson theorem says that an unbiased estimator of $\Sigma_N v_i$ is $\Sigma_N I_i v_i / p_i$, where I_i equals 1 if unit i was obtained in the sample and $I_i = 0$ otherwise. The sum in the Horvitz-Thompson theorem is over both sampled and unsampled units because $I_i = 0$ for all unsampled units. All terms in the first summation of (9.9) are obtained in the sample from animals captured with probability p_{ij}. All terms in the second summation of (9.9) are obtained in the sample with probability $p_{ij}p_{i'j}$ because capture of animal i is assumed to be independent of the capture of animals i'. Applying the Horvitz-Thompson theorem and substituting unbiased estimates for the p's and variances, we obtain the approximately unbiased estimator,

$$
\hat{var}(\hat{N}_j) = \sum_{i=1}^{n} \frac{h_{ij}(1-\hat{p}_{ij})}{\hat{p}_{ij}^2}\left[1 + \frac{3\hat{var}(\hat{p}_{ij})}{\hat{p}_{ij}^2} + \frac{[\hat{var}(\hat{p}_{ij})]^2}{\hat{p}_{ij}^4}\right]
$$
$$
+ \sum_{i=1}^{n}\sum_{i'=1}^{n} \frac{h_{ij}h_{i'j}\hat{cov}(\hat{p}_{ij}, \hat{p}_{i'j})}{\hat{p}_{ij}^2\hat{p}_{i'j}^2}
$$

(9.10)

where h_{ij} is the jth entry in the capture history for animal i, n is the total number of animals captured, $\hat{var}(\hat{p}_{ij})$ is an estimate of the capture probability variance for the ith individual at occasion j, and $\hat{cov}(\hat{p}_{ij}, \hat{p}_{ij})$ is an estimate of the covariance in estimated capture probability between animal i and animal i' at occasion j. Using a logistic link function, we can estimate both $\hat{var}(\hat{p}_{ij})$ and $\hat{cov}(\hat{p}_{ij}, \hat{p}_{i'j})$ from (9.7) since $\hat{var}(\hat{p}_{ij}) = \hat{cov}(\hat{p}_{ij}, \hat{p}_{ij})$.

If we define the coefficient of variation (CV) for the estimated capture probabilities as $\hat{cv}(\hat{p}_{ij}) = \sqrt{\hat{var}(\hat{p}_{ij})}/\hat{p}_{ij}$, the bracketed term in (9.10) becomes $1 + 3\hat{cv}(\hat{p}_{ij})^2 + \hat{cv}(\hat{p}_{ij})^4$. This term is a 4th-degree polynomial in $\hat{cv}(\hat{p}_{ij})$ and, along with $\hat{cov}(\hat{p}_{ij}, \hat{p}_{i'j})$, represents a penalty in variance to be paid for imprecise estimation of capture probabilities. The CV term is small (<1.28) if the CV of estimated capture probability is less than 30%, but is substantial (>5) if CV is greater than 1.0. In cases where capture probabilities are precisely estimated and most CVs are small, it will be reasonable to drop this term. If dropped, the estimator in (9.10) is equivalent to the variance estimators of Huggins (1989), Borchers et al. (1998), and Taylor et al. (2002). McDonald and Amstrup (2001) derived a different variance estimator along these same lines using 1st-order Taylor series approximations, but without the covariance terms. While their estimator is easy to calculate by hand, and is reasonably accurate when capture heterogeneity is small, it will underestimate the variance of population size in many cases. If possible, the estimator in (9.10) should be favored over that of McDonald and Amstrup (2001).

Equation 9.10 is complicated and difficult to compute. Fortunately, both 9.10 and 9.10 (without the CV term) are computed by the MRAWIN software package. Running the fully time-dependent CJS model by sex on the dipper data (table 9.1), and requesting that F.cr.estim estimate variances by inverting the Hessian of the maximization, produces the same parameter estimates as in Table 9.7, but slightly different standard error estimates. For male birds, the standard errors estimated by F.cr.estim for capture probabilities were 0.1980, 0.0000, 0.0822, 0.0612, and 0.0601 for $\hat{p}_2$ through $\hat{p}_6$, respectively. For females, estimated standard errors were 0.2037, 0.1302, 0.0703, 0.0748, and 0.0567 for $\hat{p}_2$ through $\hat{p}_6$, respectively. Applying (9.10) to this model resulted in the estimated standard errors $se(\hat{N}_2) = 20.23$, $se(\hat{N}_3) = 7.77$, $se(\hat{N}_4) = 6.30$, $se(\hat{N}_5) = 6.39$ and $se(\hat{N}_6) = 5.86$. Because the estimated CVs of capture probabilities were all <30%, it might be reasonable to apply (9.10) without the CV term. Doing so resulted in the estimated standard errors $se(\hat{N}_2) = 19.97$, $se(\hat{N}_3) = 7.74$, $se(\hat{N}_4) = 6.29$, $se(\hat{N}_5) = 6.37$, and $se(\hat{N}_6) = 5.85$. Even though it remains small, the largest difference between these two sets of variance estimates occurs at occasion 2 where CVs are highest.

9.6 A Multistate (Multistrata) Model

In this section, we use a modified version of the dipper data to demonstrate a multistate model in MARK. Whereas these models were originally conceived to explain movements of animals among geographic strata, they can explain any change of an animal's state, as discussed in chapter 8. Multistate models are generalizations of the CJS model, which allow for the movement of animals between states. The state of an animal at each sampling occasion may refer to a geographic location, a physiological state (such as nesting), or any other condition of interest. Data for multistate models are similar to conventional capture–recapture data. Each animal is represented by an individual capture history indicating whether or not it was encountered at each occasion (e.g., a series of 0's and 1's for CJS studies). For multistate models, however, the generic 1 used to designate a capture is replaced by a variable that reflects the state in which the animal was encountered. For example, the multistate capture history "AB0B" describes an animal that was caught at time 1 in state A, was caught at time 2 in state B, was not caught at time 3, and was last caught at time 4 in state B.

For the purposes of this example, the dipper data set was transformed into an artificial multistate data set by randomly assigning each capture to either state A or state B by flipping a coin. No distinction is made between the two groups (male and female) for this example, and therefore each

individual capture history is followed by a single 1 to represent its frequency. The first 5 lines of the 294 modified individual capture histories
look like

BAAAAA0 1

BABAB00 1

BBBB000 1

BAAB000 1

AA0BAB0 1

Because the designation of states was random, there are roughly as
many captures in state A as in state B at each occasion, and we therefore expect that the probabilities of transition between the two states
(ψ^{AB} and ψ^{BA}) estimated by MARK will be close to 0.5. The modified
dipper data set was read into MARK specifying the *Multi-strata Recaptures only* data type, 7 encounter occasions, 1 attribute group, and 2
strata.

The three fundamental parameters in a multistate model are defined as
follows: S_i^A is the probability that an animal in state A at sampling occasion i survives and remains in the study population until sampling occasion $i + 1$, p_i^A is the probability that an animal alive in state A at time i is
captured, and ψ_i^{AB} is the probability that an animal is in state B at sampling occasion $i + 1$, given that the animal was in state A at period i and
that it survived until $i + 1$ and remained in the study population. To illustrate the mechanics of running a multistate model in MARK, we consider
a simple model which assumes that the probabilities of survival, capture,
and transition are constant within each state for the duration of the study,
but are different between the two states. An appropriate name for this
model is ($S^s p^s \psi^s$), where the superscript s denotes the fact that parameters
are allowed to vary by state. Model ($S^s p^s \psi^s$) has 6 independent parame-
ters: S^A, S^B, P^A, P^B, ψ^{AB}, and ψ^{BA}. The structure of these parameters is represented by 6 PIMs in program MARK. To specify model ($S^s p^s \psi^s$), a sin-
gle index value is entered into each of the 6 PIMs. For example, all values
in the survival PIM for state A are set to 1, all values in the survival PIM
for state B are set to 2, all values in the capture PIM for state A are set to 3,
and so on. The structures of all of the PIMs for model ($S^s p^s \psi^s$) are shown
in table 9.17. The reader should be aware that, as with other types of
capture–recapture modeling in program MARK, it is possible to fit more
complex multistate models through manipulation of the design matrix.
For this example, however, we keep things simple by running a model that
is most easily specified through the PIMs.

TABLE 9.17
The MARK PIMs for the European dipper multistate model ($S^s\ p^s\ \psi^s$)

Survival parameter (S) for state A

1	1	1	1	1	1
	1	1	1	1	1
		1	1	1	1
			1	1	1
				1	1
					1

Survival parameter (S) for state B

2	2	2	2	2	2
	2	2	2	2	2
		2	2	2	2
			2	2	2
				2	2
					2

Capture parameter (p) for state A

3	3	3	3	3	3
	3	3	3	3	3
		3	3	3	3
			3	3	3
				3	3
					3

Capture parameter (p) for state B

4	4	4	4	4	4
	4	4	4	4	4
		4	4	4	4
			4	4	4
				4	4
					4

Transition parameter (Psi) for state A

5	5	5	5	5	5
	5	5	5	5	5
		5	5	5	5
			5	5	5
				5	5
					5

Transition parameter (Psi) for state B

6	6	6	6	6	6
	6	6	6	6	6
		6	6	6	6
			6	6	6
				6	6
					6

Running model ($S^s p^s \psi^s$) in MARK results in a model deviance of 227.17 and the estimates $S^A = 0.557$, $S^B = 0.564$, $p^A = 0.889$, $p^B = 0.920$, $\psi^{AB} = 0.471$, and $\psi^{BA} = 0.601$. Recall that these estimates depend on the random assignment of dipper captures to states A and B, and will therefore vary slightly upon repetition. As expected, the probabilities of transition between the two states in this example are close to 0.5.

9.7 Polar Bears in the Southern Beaufort Sea

The classic model for estimating abundance of open populations is the Jolly-Seber (JS) model (Pollock et al. 1990). Failure of the assumptions of homogeneity of survival and capture probabilities, however, has essentially prevented use of the JS model in many wildlife studies. The case of polar bears of the Southern Beaufort Sea (SBS) is no exception. There are many potential causes of heterogeneity in polar bear capture probabilities. Polar bears can live up to 30 years in the wild (DeMaster and Stirling

1981; Amstrup 2003), and are more individualistic than innate in their behaviors. Capture probabilities for individual bears, therefore, may vary greatly. Capture effort and success has varied among years because of natural factors like weather and sea ice, and because of changing directives from funding agencies. Further, radiotelemetry data show that individual bears may not be uniformly available for capture. Here, we show how to account for the heterogeneity in both capture and survival probabilities of polar bears by relating those histories to covariates using the general regression approach described above. A Horvitz-Thompson-type (section 9.5) estimator is then used to estimate population size.

Female polar bears in the SBS were captured by drug immobilization during spring and autumn field seasons. Because mortality among polar bears is believed to be minimal during the summer (Amstrup et al. 2001), the spring and autumn capture events were pooled into single annual occasions. The data set consisted of 32 capture occasions from 1967 to 1998, during which 1025 female polar bears were recaptured 1617 times (Amstrup et al. 2001). Three of the 1025 lines of data follow (note the 2 in line one denotes death on capture):

$$00001101010001002000000000000000$$
$$00101011000010001000000000000000$$
$$00000000000000000000000001100101$$

The fundamental parameters ϕ_{ij} and p_{ij} were modeled as a function of covariates that (1) changed through time but were constant among animals, (2) changed among animals but were constant through time, and (3) changed through time and among animals. Modeling ϕ_{ij} and p_{ij} as functions of covariates alleviated the need to make assumptions about survival and capture probabilities.

In conceptualizing this as a regression problem as discussed above, covariates were thought of as two-dimensional arrays with rows representing individual animals and columns representing capture occasions, and the models for ϕ_{ij} and p_{ij} were conceptualized and written like equations 9.3 and 9.4. Once the coefficients β_a and γ_a were estimated, estimates of survival and capture probabilities were computed as

$$\hat{\phi}_{ij} = \frac{e^{\hat{\beta}_0 + \hat{\beta}_1 x_{ij1} + \hat{\beta}_2 x_{ij2} + \cdots + \hat{\beta}_p x_{ijp}}}{1 + e^{\hat{\beta}_0 + \hat{\beta}_1 x_{ij1} + \hat{\beta}_2 x_{ij2} + \cdots + \hat{\beta}_p x_{ijp}}}$$

and

$$\hat{p}_{ij} = \frac{e^{\hat{\gamma}_0 + \hat{\gamma}_1 z_{ij1} + \hat{\gamma}_2 z_{ij2} + \cdots + \hat{\gamma}_q z_{ijq}}}{1 + e^{\hat{\gamma}_0 + \hat{\gamma}_1 z_{ij1} + \hat{\gamma}_2 z_{ij2} + \cdots + \hat{\gamma}_q z_{ijq}}}$$

Maximization of the CJS likelihood (chapter 5) to obtain coefficient estimates under this parameterization was carried out using the S-Plus or R routines described above (www.west-inc.com). Once coefficients in the regression equations and the survival and capture probabilities were estimated, population sizes were estimated using the Horvitz-Thompson estimator of equation 5.5.

We began the model building process with a list of biologically relevant covariates that had been recorded (table 9.18). Most of the covariates considered fit logically into survival or capture equations, but not both. For example, capture effort (helicopter flight hours) and geographic region of capture (latitude and longitude) seemed unlikely to explain variation in survival probability. Both, however, were logical contributors to variation in capture probability. Candidate models containing many, but not all, combinations of covariates were ranked by three likelihood criteria: (1) the drop in deviance, (2) AIC, and (3) QAIC, the quasi-likelihood adjustment to AIC that relies on an estimate of the overdispersion $\hat{c}$. Here, we demonstrate computation of the coefficient values, parameters of interest, and deviance, and AIC values for the model that seemed most sensible for this data set. Note that subsequent to the original publication of this approach, we updated our version of Program S+. The newer version exports more digits than the previous, and, not surprisingly, there are minor differences between the results shown here those reported earlier (Amstrup et al. 2001; McDonald and Amstrup 2001).

After model selection that involved trade-offs between model fit and variance of the population size estimates (Amstrup et al. 2001), the best approximating model for the data was determined to be

$$\ln\left(\frac{p_{ij}}{1 - p_{ij}}\right) = -1.863 + 6.4813(radio) + 0.0021(west.effort)$$

$$+ 0.0046(east.effort) - 0.0848(year) - 1.3707(region)$$
$$+ 0.0035(region \times west.effort) - 0.003(region \times east.effort)$$

and

$$\ln\left(\frac{\phi_{ij}}{1 - \phi_{ij}}\right) = 0.6744 + 5.1909(sigmoidal.year) + 0.6116(age.class.2)$$

where *age.class.2* is the indicator for the 2nd age class only, and the covariates are defined in table 9.18.

To illustrate the computations, the estimated capture probability from the best approximating model for a noninstrumented eastern polar bear

TABLE 9.18

The variables considered in the polar bear study in the Southern Beaufort Sea

Covariate	Description	Type
Year	Linear trend though time. Coded as 1 through 32 corresponding to each year.	Time
Year.squared	Year*Year; facilitated fitting of a quadratic effect.	Time
Sigmoidal.year	Year transformed through a logistic function. i.e., $sigmoidal.year = 1 - [1 + \exp(0.5x)]^{-1}$, where $x = (year - 14.5)$. E.g., for 1982, $x = 16 - 14.5$, and $sigmoidal.year = 1 - 1/\{1 + \exp[0.5(1.5)]\} = 0.6792$.	Time
Linear age	Age at first encounter, incremented each year for surviving bears	Individual and time
Age.class	Age grouped as (1) 0 to 4 years old, (2) 5 to 20 years old, (3) 21 years and older. Animal gets a 1 for the age group it is in and a 0 otherwise.	Individual and time
Harvest	Yearly total number of bears harvested in the Alaskan and Canadian Beaufort Sea.	Time
Ice	Indicator of the severity of winter, coded as 0, normal ice pack; and 1, heavy ice pack.	Time
Radio	Indicator for whether a bear was wearing radio collar at each capture occasion, coded as 0, bear not wearing radio collar during occasion; and 1, bear wearing radio collar during occasion.	Individual and time
Effort	Yearly total number of flight hours spent searching the Beaufort sea study area by American and Canadian researchers.	Time
West.Effort	Number of hours searching for polar bears flown west of 155 longitude.	Time
East.Effort	Number of hours searching for polar bears flown east of 154.99 longitude.	Time
Group size	Average number of bears in the group when bear was captured. Average is over all trap occasions where bear was seen.	Individual
Group size class	Mean group size over bear's capture history, coded as 1, group size = 1; 2, 1 < group size ≤ 2; 3, 2 < group size ≤ 3; 4, group size > 3.	Individual

TABLE 9.18 *continued*

Covariate	Description	Type
Group-size Squared	The square of Mean Group Size used because of the apparent hump in frequencies of group sizes observed.	Individual
Sightings	The mean number of sightings of a particular bear per year. This was the average number of sightings over the life of the project, calculated as Total #Sightings/#Years sighted.	Individual
Period (2 level)	Indicator for time period within the study. Coded as 1, 1967 to 1980; 2, 1981 to 1994.	Time
Period (4 level)	Indicator for time period within the study. Coded as 1, 1967 to 1973; 2, 1974 to 1980; 3, 1981 to 1987; 4, 1988 to 1994.	Time
Region	Indicator for region in which the bear generally lived. If the harmonic mean center of activity (COA) derived from all of an individual's relocations was east of 155 longitude, the bear was assigned to the "east" = 0 region (Southern Beaufort Sea); otherwise it was assigned to the "west" = 1 region (Eastern Chuckchi Sea).	Individual

Note. In addition, certain two-way interactions were considered. Type equal to "Time" means the variable varied across trap occasions (years in this case) but not across individuals. Type equal to "Individual" means the variable varied across individuals but not trap occasion. Type equal to "Individual and time" means the variable varied across trap occasions and individuals.

in the 15th year of the study (1981) when west effort = 20 hours and east effort = 100 hours was

$$\ln\left(\frac{p_{ij}}{1-p_{ij}}\right) = -1.863 + 6.4813(0) + 0.0021(20) + 0.0046(100) - 0.0848(15)$$

$$-1.3707(0) + 0.0035(0)(20) - 0.003(0)(100)$$

the right-hand side of which reduces to $-1.863 + 0.042 + 0.46 - 1.272 = -2.633$. Solving this equation gives $\hat{p}_{ij} = e^{-2.633}/(1 + e^{-2.633}) = 0.0718/1.0718 = 0.06702$.

A radio-collared eastern bear in the same year would have had a capture probability (replace 6.4813(0) with 6.4813(1) in the previous calculation) of

$$\hat{p}_{ij} = \frac{e^{3.85}}{1 + e^{3.85}} = 0.98$$

reflecting the fact that instrumented bears were almost always relocated at least once per year. This calculation illustrates the way in which capture histories from radio-collared animals were included along with other histories and the resulting heterogeneity modeled with a covariate (i.e., radio).

For a noninstrumented western polar bear captured in the 15th year of study when west effort = 20 hours and east effort = 100 hours, estimated capture probability was

$$\ln\left(\frac{p_{ij}}{1 - p_{ij}}\right) = -1.863 + 6.4813(0) + 0.0021(20) + 0.0046(100) - 0.0848(15)$$

$$-1.3707(0) + 0.0035(0)(20) - 0.003(1)(100)$$

This reduces to $-1.863 + 0.042 + 0.46 - 1.272 - 1.3707 + 0.07 - 0.3 = -4.2337$. Hence,

$$\hat{p}_{ij} = \frac{e^{-4.2337}}{1 + e^{-4.2337}} = \frac{0.01450}{1.01450} = 0.0143$$

Similarly, the estimated survival rate for a 15-year-old polar bear captured in the 15th year of the project would be

$$\ln\left(\frac{\phi_{ij}}{1 - \phi_{ij}}\right) = 0.6744 + 5.1909(0.56218) + 0.6116(1) = 4.20$$

so that

$$\phi_{ij} = \frac{e^{4.2}}{1 + e^{4.2}} = \frac{66.69}{67.69} = 0.985$$

Figure 9.4 shows estimates of $\hat{N}_j$ provided by the best approximating model. As explained in chapter 5 and earlier in this chapter, the Horvitz-Thompson estimator of $\hat{N}_j$ is the sum of the reciprocals of the capture probabilities for animals captured in each year. Considering the probabilities just calculated, each noninstrumented eastern bear captured in year

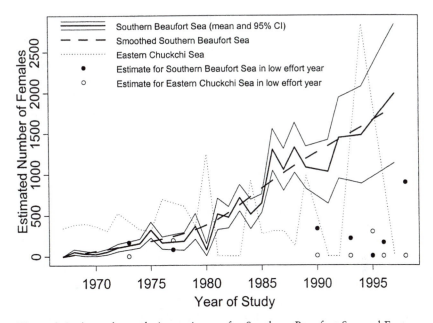

Figure 9.4. Annual population estimates for Southern Beaufort Sea and Eastern Chuckchi Sea polar bears from the best approximating capture–recapture model. Absence of capture efforts caused $\hat{N}_j$ to dip in 1973, 1977, 1990, 1993, 1995, and 1996. also, in 1998, capture efforts were only 1/10 of typical-pushing that estimate down in a similar fashion. Spikes in $\hat{N}_j$ for the Eastern Chuckchi during 1980, 1984, 1989, and 1994 were caused by greater than usual representation of bears with very small capture probabilities.

15 contributed 1/0.06702 or 14.92 bears to the year 15 estimate. Noninstrumented western bears captured in year 15 each contributed 1/0.0143 or 69.93 bears. Note that each radio-collared bear contributed only 1/0.98 or 1.02 bears to the total estimate for the 15th year of the study.

Smoothing the annual estimates from this model indicates that the female polar bear population could have been as high as 1800 in the late 1990s. Note that years of no or minimal capture effort and few captures push the Horvitz-Thompson estimate to very low levels and increase the estimated standard errors. Estimates derived for those years (1973, 1977, 1990, 1993, 1995, and 1996) were not included in the smoothed estimates of $\hat{N}_j$ shown in figure 9.4, nor were they included in calculations of CVs discussed below. Assuming a population that is 60% female, the resulting late 1990s estimate of total Southern Beaufort Sea population size (males and females) could have been 3000. A more even sex ratio, of course, would translate into a larger total population. The population estimates derived from this analysis are consistent with empirical evidence of a population increase (Amstrup 2000; Stirling and Andriashek 1992) and associated with the narrowest interval estimates to date for this population. CVs on $\hat{N_j}$ ranged from 0.09 to 0.48 over all years of the study (except those noted above), and averaged of 0.17 between 1988 and 1997. These CVs compare favorably with the mean CV of 0.145 on female $\hat{N_j}$ in Hudson Bay (Lunn et al. 1997), where a much higher and more consistent portion of the population was marked.

Figure 9.3 does not show the estimate of size for 1998. Estimates of capture probability for the last study occasion are unreliable in the CJS estimation format (chapters 3 and 5). The general regression approach that utilizes covariates, combined with the Horvitz-Thompson estimator, can provide an estimate for the last occasion, but its reliability is thought to be poor. Capture effort was extremely low in 1998, forcing the smoothed estimates of $\hat{N_j}$ downward (Amstrup et al. 2001). Leaving 1998 out of the smoothing results in an apparently strongly increasing population in the late 1990s. We still urge caution in interpretation of the trends in numbers reported here. Error estimates shown here are known to be too small in the presence of high capture heterogeneity, which is common among polar bears (McDonald and Amstrup 2001), and in the last few years of the study, the CVs were very high·

9.8 Dead Recoveries of Mallard Ducks

In this example, we illustrate analyses of a dead-recovery data set for adult female mallards (*Anas platyrhynchos*) using program MARK. The data set used in this example was originally presented by Nichols et al. (1982a) and was later reanalyzed by Williams et al. (2001). The data were collected from 1966 to 1978 in Canada with 13 sampling occasions. Wild mallards were banded during July through September for 13 consecutive years, and bands were found or returned by hunters in September through February of each year. Over 29,000 birds were banded in total. This is an example of a single-age, single-group, dead-recovery data set including one environmental covariate. The most obvious difference between this and the preceding examples with the dipper data is the concept of dead rather than live recoveries.

As in all types of capture–recapture, the analysis of dead-recovery data should be guided by the underlying biology and study design. Mallard population dynamics have been shown to vary over time in response to environmental conditions (Nichols et al. 1982a; Anderson 1975), and Nichols et al. (1982a) suggested that the natural mortality rates of mallards in this study were correlated with the densities of mallards on ponds in the prairie breeding areas. Harvest, retrieval, and reporting rates were expected to vary over the course of the study as a function of hunting regulations and other factors. A temporary banding effect, which allows for different probabilities of being shot and recovered in the first year after banding, was also suspected. This is a common condition in waterfowl studies when a large number of banded birds are harvested before they leave the banding site.

Dead-recovery data are generally presented either as a list of individual encounter histories or in a recovery matrix. Encounter histories for the 13 sampling occasions comprising the mallard data take the form

$$11\ 00\ 00\ 00\ 00\ 00\ 00\ 00\ 00\ 00\ 00\ 00\ 00$$

$$10\ 01\ 00\ 00\ 00\ 00\ 00\ 00\ 00\ 00\ 00\ 00\ 00$$

$$00\ 10\ 01\ 00\ 00\ 00\ 00\ 00\ 00\ 00\ 00\ 00\ 00$$

This format is different from conventional live-recapture encounter histories, which require only one number per occasion to indicate whether an animal was seen (1) or not seen (0). Here, each pair of numbers indicates whether the bird was banded or recovered dead, or banded and recovered dead during a single interval $i = 1, \ldots, l$, where l is the total number of recovery intervals in the study. For example, the first encounter history represents a bird that was both banded and recovered dead in interval $i = 1$, the second encounter history represents a bird that was banded in $i = 1$ and recovered dead in $i = 2$, and the third encounter history represents a bird that was banded in $i = 2$ and recovered dead in $i = 3$.

Table 9.19 shows the mallard data set recovery matrix as formatted by Williams et al. (2002), while table 9.20 presents the recovery matrix as modified for input into MARK. Note that the numbers of animals marked in each occasion simply shift to the bottom row of the table as formatted for MARK.

The numbers in column 1 of table 9.19 and the numbers in the last row at the bottom of table 9.20 indicate how many birds were banded and released at each banding occasion $i = 1, \ldots, k$, where k is the total number of banding occasions in the study. The numbers in the triangular matrix indicate how many birds banded at each occasion i were subsequently recovered (reported by hunters or found dead) in recovery intervals $j = 1, \ldots, l$. For example, matrix element $m_{15} = 7$ indicates that 7 of the 926 birds banded in the first year of the study ($i = 1$) were recovered in the fifth year of the study ($j = 5$).

As with the dipper data, begin the analysis in MARK by opening and naming a new project. Select the *Brownie et al. Recoveries* radio button on the left of the *Enter Specifications for MARK Analysis* window. Set

TABLE 9.19

Brownie et al. (1985) format tag-recovery data for 29,000 mallards tagged over a 13-year period

I	R_i	$j=1$	2	3	4	5	6	7	8	9	10	11	12	$l=13$
						m_{ij}								
1	926	39	20	6	3	7	2	2	1	0	0	0	0	0
2	1413		53	18	17	9	8	5	4	0	0	1	0	0
3	1147			34	27	23	11	5	3	0	2	1	0	0
4	1233				57	33	21	14	5	2	1	2	0	1
5	1674					82	35	22	8	2	4	3	1	1
6	1727						71	30	13	1	8	6	2	0
7	1864							67	32	12	12	4	4	3
8	1438								46	23	21	13	5	4
9	1235									43	27	15	10	7
10	2351										67	56	26	13
11	5215											180	91	56
12	5256												167	89
$k=13$	3615													114

Note. Adult female mallards banded preseason 1966–1978 in Manitoba, Saskatchewan, and Eastern Alberta. R_i is the number banded and released at each occasion $i = 1, \ldots, k$. The triangular matrix indicates how many birds banded at each occasion i were reported in recovery intervals j (e.g., $m_{15} = 7$ indicates that 7 of the 926 birds banded in the first year of the study ($i = 1$) were recovered in the fifth year of the study ($j = 5$).

the number of encounter occasions to 13, and select the input text file with extension .INP. The .INP text file contains the recovery matrix and is formatted like table 9.20. In this example, we follow the parameterization scheme of Brownie et al. (1985) so that S_i is the probability of an individual surviving from time $i - 1$ to time i, and, f_i as the probability of an individual being recovered (i.e., harvested, retrieved, and reported) between time i and $i + 1$.

Once the mallard data are read into MARK, a window appears with a parameter index matrix (PIM) for the survival parameters (S_i). The PIM that appears looks like table 9.21. The numbers 1–12 in the PIM are indices that MARK will use for the survival parameters. The fact that each column contains a different number indicates that, under the current model, MARK will allow S_i to assume a different value for each year of the study. Now open the PIM for the recovery parameter (f_i) by selecting "Open Parameter Index Matrix" from the *PIM* drop-down menu on the main toolbar, clicking the *Select All* button on the right, and clicking "OK". Again, each column of the PIM contains a different number. This means that the default model in MARK is single-age (meaning all animals are treated the same regardless of age), single-group (there are no group

TABLE 9.20
Recovery matrix for mallard data, formatted for input into program MARK

```
/*Adult female mallard 1966–1978
  Data from Williams et al. (2002)
  Groups = 1
  Occasions = 13
  No individual covariates*/
  recovery matrix group = 1;
  39   20    6    3    7    2    2    1    0    0    0    0    0 ;
       53   18   17    9    8    5    4    0    0    1    0    0 ;
            34   27   23   11    5    3    0    2    1    0    0 ;
                 57   33   21   14    5    2    1    2    0    1 ;
                      82   35   22    8    2    4    3    1    1 ;
                           71   30   13    1    8    6    2    0 ;
                                67   32   12   12    4    4    3 ;
                                     46   23   21   13    5    4 ;
                                          43   27   15   10    7 ;
                                               67   56   26   13 ;
                                                   180   91   56 ;
                                                        167   89 ;
                                                             114 ;

 926  1413  1147  1233  1674  1727  1864  1438  1235  2351  5215  5256  3615 ;
```

Note. Adult female mallards banded preseason 1966–1978 in Manitoba, Saskatchewan, and Eastern Alberta.

differences due to sex, flock, or herd, etc.), and fully time-dependent (both survival and recovery values are assumed to vary among time intervals). Following Lebreton et al. (1992) and chapters 6 and 7, this model is denoted (S_t, f_t). Do not change anything in either PIM at this point because the default model in MARK provides an appropriate starting point. Select "Current Model" from the *Run* drop-down menu on the main toolbar. Name model (S_t, f_t) in the *Setup Numerical Estimation Run* window and click "OK to Run" without changing any of the default selections.

The *Results Browser* window in MARK now displays a single row with the name of the current model and six other columns of information, which are the basis for comparison between models and will be considered later. Results (parameter estimates, etc.) from the current model can be accessed through the buttons at the top of the *Results Browser* window.

A reasonable global model for the mallard data allows for variation in S_i and f_i by time and includes a temporary banding effect, which lets f_i in the first year after banding be different than in subsequent years. This model, denoted (S_t, f_t^*), is constructed by modifying the recovery parameter PIM as shown in table 9.22. Name this model and run it, as before.

TABLE 9.21

The MARK parameter information matrix (PIM) for the survival parameter (S) of the Brownie model analysis of the mallard band-recovery data

Cohort	Capture occasion											
	2	3	4	5	6	7	8	9	10	11	12	13
1	1	2	3	4	5	6	7	8	9	10	11	12
2		2	3	4	5	6	7	8	9	10	11	12
3			3	4	5	6	7	8	9	10	11	12
4				4	5	6	7	8	9	10	11	12
5					5	6	7	8	9	10	11	12
6						6	7	8	9	10	11	12
7							7	8	9	10	11	12
8								8	9	10	11	12
9									9	10	11	12
10										10	11	12
11											11	12
12												12

Note. Cohort denotes the sample of birds captured and banded during each year. Capture occasions were years in this example.

TABLE 9.22

The MARK parameter information matrix (PIM) for the recovery parameter (f) of the Brownie dead-recovery model fit to mallard band-recovery data

Cohort	Occasion												
	1	2	3	4	5	6	7	8	9	10	11	12	13
1	25	13	14	15	16	17	18	19	20	21	22	23	24
2		26	14	15	16	17	18	19	20	21	22	23	24
3			27	15	16	17	18	19	20	21	22	23	24
4				28	16	17	18	19	20	21	22	23	24
5					29	17	18	19	20	21	22	23	24
6						30	18	19	20	21	22	23	24
7							31	19	20	21	22	23	24
8								32	20	21	22	23	24
9									33	21	22	23	24
10										34	22	23	24
11											35	23	24
12												36	24
13													37

Note. Cohort denotes the sample of birds captured and banded during each year. Capture occasions were annual in this example.

The AIC selection criteria that will be used to compare between models assume that at least one of the models under consideration is an adequate representation of the data (Burnham and Anderson 1998). Therefore, the goodness of fit of the global model should be tested before continuing with model construction. Goodness-of-fit testing by parametric bootstrapping in MARK requires a temporary switch to an alternate parameterization, denoted (S, r), as discussed in chapter 10 of Cooch and White (2001). Fortunately, MARK allows a change of the parameterization scheme mid-analysis. Highlight model (S_t, f_t^*) in the *Results Browser* window, select "Change Data Type" from the *PIM* drop-down menu on the main toolbar, and select the "Dead Recoveries" data type in the window that appears. Now rerun the model with name (S_t, r_t^*). Note that the new parameterization does not change the logic behind the model or the way it fits the data, as confirmed by identical results for models (S_t, f_t^*) and (S_t, r_t^*) in the *Results Browser* window. Select "Bootstrap GOF" from the *Tests* drop-down menu on the main toolbar, designate an output file, and run between 500 and 1000 simulations. Once completed, select "View Simulation results" from the *Simulations* drop-down menu on the main toolbar and open the output file. The simulation results can be sorted according to deviance by clicking the 3rd button from the right in the *Browse Database* window. Recall from the *Results Browser* window that the deviance of model (S_t, r_t^*) was 35.9. Comparison of this value to the simulation results reveals that most of the simulations have appreciably higher deviances than the actual model. The data sets from which the simulation deviances were derived were created to fit the selected model and satisfy all assumptions. If the global model were a poor fit to the data, we would expect the deviances of the simulations to be smaller than that of the global model. Because most of the deviances from the simulation data sets exceeded that of the global model, it appears that the global model does indeed adequately fit the data.

A reasonable list of restricted models can now be developed for the mallard data. Comparison between these models will reveal which biological effects are supported by the data, thus providing a means of hypothesis testing. Less parameterized models may have higher estimated precision than the global model, but also may show an increased level of bias. AIC and its derivatives provides an objective method for evaluating this trade-off and selecting the best model.

The mallard analysis began with the MARK default model, (S_t, f_t), and the global model (S_t, f_t^*), which allowed for a banding effect in addition to full time dependence. A logical next step is to investigate the significance of time dependence in each parameter separately by building three additional models—$(S_t, f_.)$, $(S_., f_t)$, and $(S_., f_.)$. As before, these

TABLE 9.23

The MARK parameter information matrices (PIMs) for the constant (homogenous) model (S, f) in the Brownie model of for the mallard band-recovery data

	PIM for Survival Parameter (S)											
						Occasion						
Cohort	2	3	4	5	6	7	8	9	10	11	12	13
1	1	1	1	1	1	1	1	1	1	1	1	1
2		1	1	1	1	1	1	1	1	1	1	1
3			1	1	1	1	1	1	1	1	1	1
4				1	1	1	1	1	1	1	1	1
5					1	1	1	1	1	1	1	1
6						1	1	1	1	1	1	1
7							1	1	1	1	1	1
8								1	1	1	1	1
9									1	1	1	1
10										1	1	1
11											1	1
12												1

	PIM for Recovery Parameter (f)												
						Occasion							
Cohort	1	2	3	4	5	6	7	8	9	10	11	12	13
1	2	2	2	2	2	2	2	2	2	2	2	2	2
2		2	2	2	2	2	2	2	2	2	2	2	2
3			2	2	2	2	2	2	2	2	2	2	2
4				2	2	2	2	2	2	2	2	2	2
5					2	2	2	2	2	2	2	2	2
6						2	2	2	2	2	2	2	2
7							2	2	2	2	2	2	2
8								2	2	2	2	2	2
9									2	2	2	2	2
10										2	2	2	2
11											2	2	2
12												2	2
13													2

models are created through the PIMs. For example, the notation (S, f) corresponds to a model that restricts both S_i and f_i to be constant for the duration of the study. Both PIMs for this model should therefore contain the same index value for all rows and columns (recall that PIM columns correspond to study years, as shown in table 9.23).

Run models (S_t, f), $(S_., f_t)$, and $(S_., f_.)$ and compare the output in the *Results Browser* window, which should include all 5 models considered thus far. Ranking the models by AIC_c shows, for example, a ΔAIC_c difference of 29.78 between models $(S_., f_t)$ and $(S_., f_.)$, which indicates that the data show strong support for time variation in the reporting rate.

At this point, additional biological information is required to proceed with the analysis. As mentioned at the beginning of the example, Nichols et al. (1982a) recognized a negative density dependence in the natural mortality rate of mallards. The significance of this effect can be investigated by modeling survival as a function of the yearly covariate *ducks/pond*, for which Williams et al. (2002) provided standardized values. In MARK, the modeling of environmental covariates is accomplished through the Design Matrix. See chapter 7 of Cooch and White (2001) for more discussion of the Design Matrix and linear models. For now, retrieve model (S_t, f_t) and select "Full" from the *Design* drop-down menu on the main toolbar. The corresponding Design Matrix represents the structure of the linear models underlying the time-dependent model.

The goal of modeling the environmental covariate *ducks/pond* is to add independent information to the analysis, thereby explaining some of the variation in S_i without incurring the penalty in AIC_c associated with the large number of parameters required for full time dependence. In other words, whereas model (S_t, f_t) models survival as a function of 12 coefficients (B1–B12), an alternate covariate model constrains S_i to be a linear function of *ducks/pond*, i.e.,

$$\ln\left(\frac{S_i}{1 - S_i}\right) = \beta_0 + \beta_1 x_i$$

where x_i is the value of *ducks/pond* for year i, and β_0 and β_1 are estimated coefficients. This model can be specified in MARK by deleting columns B3–B12 of the time-dependent Design Matrix and entering each year's value of the covariate *ducks/pond* into column B2. The resulting Design Matrix represents the model (S_{cov}, f_t) and is shown in table 9.24. Name and run this model.

The *Results Browser* window now includes all models developed to this point. AIC_c values indicate that (S_{cov}, f_t) is the most useful representation of the mallard data because model (S_{cov}, f_t) has an appreciably lower deviance than model $(S_., f_t)$ and requires only 1 more parameter. It can therefore be concluded that survival rates of the adult female mallards in this study were predicted by population density on spring breeding grounds (as represented by *ducks/pond*), and that the band reporting

TABLE 9.24

The MARK design matrix for the reduced-parameter, Brownie et al. model (S_{cov}, f_t) fit to the mallard band-recovery data

B1	B2	Param	B3	B4	B5	B6	B7	B8	B9	B10	B11	B12	B13	B14	B15
1	0.978	1:S	0	0	0	0	0	0	0	0	0	0	0	0	0
1	2.245	2:S	0	0	0	0	0	0	0	0	0	0	0	0	0
1	1.268	3:S	0	0	0	0	0	0	0	0	0	0	0	0	0
1	1.090	4:S	0	0	0	0	0	0	0	0	0	0	0	0	0
1	1.336	5:S	0	0	0	0	0	0	0	0	0	0	0	0	0
1	1.307	6:S	0	0	0	0	0	0	0	0	0	0	0	0	0
1	2.260	7:S	0	0	0	0	0	0	0	0	0	0	0	0	0
1	0.665	8:S	0	0	0	0	0	0	0	0	0	0	0	0	0
1	0.890	9:S	0	0	0	0	0	0	0	0	0	0	0	0	0
1	1.318	10:S	0	0	0	0	0	0	0	0	0	0	0	0	0
1	2.116	11:S	0	0	0	0	0	0	0	0	0	0	0	0	0
1	0.871	12:S	0	0	0	0	0	0	0	0	0	0	0	0	0
0	0	13:f	1	1	0	0	0	0	0	0	0	0	0	0	0
0	0	14:f	1	0	1	0	0	0	0	0	0	0	0	0	0
0	0	15:f	1	0	0	1	0	0	0	0	0	0	0	0	0
0	0	16:f	1	0	0	0	1	0	0	0	0	0	0	0	0
0	0	17:f	1	0	0	0	0	1	0	0	0	0	0	0	0
0	0	18:f	1	0	0	0	0	0	1	0	0	0	0	0	0
0	0	19:f	1	0	0	0	0	0	0	1	0	0	0	0	0
0	0	20:f	1	0	0	0	0	0	0	0	1	0	0	0	0
0	0	21:f	1	0	0	0	0	0	0	0	0	1	0	0	0
0	0	22:f	1	0	0	0	0	0	0	0	0	0	1	0	0
0	0	23:f	1	0	0	0	0	0	0	0	0	0	0	1	0
0	0	24:f	1	0	0	0	0	0	0	0	0	0	0	0	1
0	0	25:f	1	0	0	0	0	0	0	0	0	0	0	0	0

Note. Column B2 is the covariate *ducks/pond*, which helps explain yearly variation in survival probabilities. Param indicates real parameter designations.

TABLE 9.25
Covariate (ducks/pond) and estimates of survival probabilities (S_i) and recovery probabilities (f_i) for mallard tag-recovery data, using model (S_{cov}, f_i)

Year	Ducks/ pond	Estimate (S_i)	Standard error (S_i)	95% Confidence interval (S_i) Lower	Upper	Estimate (f_i)	Standard error (f_i)	95% Confidence interval (f_i) Lower	Upper
1966	NA	NA	NA	NA	NA	0.042	0.007	0.031	0.057
1967	0.978	0.631	0.022	0.588	0.672	0.037	0.004	0.029	0.046
1968	2.245	0.489	0.035	0.420	0.558	0.027	0.004	0.021	0.036
1969	1.268	0.600	0.013	0.574	0.625	0.042	0.004	0.034	0.051
1970	1.090	0.619	0.018	0.583	0.654	0.048	0.004	0.041	0.056
1971	1.336	0.592	0.011	0.569	0.614	0.041	0.003	0.034	0.048
1972	1.307	0.595	0.012	0.572	0.619	0.036	0.003	0.030	0.042
1973	2.260	0.487	0.036	0.417	0.557	0.033	0.003	0.027	0.040
1974	0.665	0.664	0.032	0.599	0.723	0.024	0.003	0.019	0.030
1975	0.890	0.641	0.024	0.592	0.687	0.031	0.003	0.026	0.037
1976	1.318	0.594	0.012	0.571	0.617	0.035	0.002	0.031	0.040
1977	2.116	0.504	0.030	0.445	0.562	0.033	0.002	0.029	0.037
1978	0.871	0.643	0.025	0.592	0.690	0.030	0.002	0.027	0.034

rates varied from year to year. Parameter estimates for model (S_{cov}, f_t) as output by program MARK are displayed in table 9.25.

9.9 Chapter Summary

- Empirical examples of models described in earlier chapters were given.
- The Jolly-Seber model was applied to the European dipper data.
- The Manly-Parr model was applied to the European dipper data.
- Cormack-Jolly-Seber model was applied to the European dipper data. This included illustration of three ways to fit the model using program MARK (PIMs, design matrix, and individual covariate). A general regression approach was also provided that avoided PIM's and MARK's design matrix.
- Significant liberties were taken with the dipper data and the Huggins closed-population model was applied. This example included illustration of the likelihood in a spreadsheet program.

- The Test 2 and Test 3 goodness-of-fit procedures (available in program RELEASE) were discussed and illustrated. The parametric bootstrap procedure was also described and illustrated using the dipper data. Overdispersion was estimated based on Test 2 and Test 3.
- The Horvitz-Thompson estimator of population size was applied to the dipper data.
- Significant liberties were again taken with the dipper data, it was modified, and a multistate model was applied.
- An involved example of CJS modeling of polar bears in the Southern Beaufort Sea was given.
- An involved example of the Brownie tag-return model applied to mallard data was given.

Figure 9.5. A poncho-marked and leg-banded sharp-tailed grouse
(*Tympanuchus phasianellus*) ready for release near Decker, Montana, spring
1977. (Photo by Steven C. Amstrup)

Ten

Capture–Recapture Methods in Practice

BRYAN F. J. MANLY, STEVEN C. AMSTRUP,
AND TRENT L. McDONALD

10.1 Introduction

The authors of chapters 2 to 8 in this book have covered the theory and applications of capture–recapture methods from the simple two-sample, closed-population situations considered by Peterson and Lincoln, through to complex multisample, multistrata, open-population situations that can be modeled only using sophisticated computer software operating on a fast modern computer. In chapter 9, we have provided empirical examples of many methods described in earlier chapters. In this final chapter we summarize the methods that have been covered and provide some closing comments, aimed particularly at readers who are intending to use these methods for analyzing their own data for the first time.

10.2 Closed-population Models

Closed-population models are discussed by Anne Chao and Richard Huggins in chapters 2 and 4. In these chapters they present a hierarchical increase in the complexity and capability of closed-population models. They start in chapter 2 with descriptions of methodological developments before the early 1960s. These early models were limited to the analysis of two or more capture occasions, and probabilities of capture were either assumed to be constant over time or allowed to vary among sampling occasions. Chapter 2 concludes by describing the limitations of the early approaches, and provides the foundation for launching into the more advanced models of chapter 4.

Chapter 4 moves on to more recent approaches for estimation of closed populations. It begins by describing the monograph of Otis et al. (1978), wherein a set of eight models are used to account for heterogeneity in capture probabilities. Chapter 4 also covers numerous alternative approaches that have been developed since 1978 for estimating population size.

The historical context presented in chapters 2 and 4 is valuable, but the large number of analytical choices presented may be intimidating and confusing for someone needing to analyze closed population capture–recapture data for the first time. Fortunately, the method of Huggins (1989, 1991) described in chapter 4 is flexible enough to handle most closed-population estimation problems from the simple to the complex. This approach uses covariates to account for variation in capture probabilities. We recommend it as the preferred method for closed-population analyses when this variation needs to be modelled.

Our preference for the Huggins (1989, 1991) approach is based on several important factors:

1. The approach is based on maximum likelihood estimation, with the advantages that this provides in terms of variance estimation, chi-squared test for comparing the fit of different models, and the use of AIC for model selection.

2. It uses logistic regression models to relate capture probabilities to covariates, in the same way that this is often done outside the area of capture–recapture analyses. This approach is very flexible, and many biologists and managers are familiar with this type of modeling.

3. The method of using covariates is similar to the modern approach for analyzing open population capture–recapture data, as described by Nichols in chapter 5.

4. Readily available computer programs such as MARK and CARE-2 are available to do the required calculations.

Chao and Huggins conclude chapter 4 with a discussion of continuous-time capture–recapture models, where individuals can be captured at any time, rather than only at a fixed number of discrete sampling times. These continuous-time models have received less attention than the discrete-time models, but may be used more in future. Lack of computer programs for carrying out the calculations required restricts their usefulness at the present time.

10.3 Open-population Models

Chapters 3 and 5 refer specifically to situations where k samples are taken from a population for which the size may vary from one sampling occasion to the next because of one or more of births, deaths, immigration, and permanent emigration. As this is the situation with most real populations, models for estimating parameters of open populations have received a great deal of attention for more than 50 years.

In chapter 3 Kenneth Pollock and Russell Alpizar-Jara review some of the early developments in open-population modeling, and then move on to the important Jolly-Seber (JS) model for estimating population sizes, survival rates, birth numbers, and capture probabilities. That discussion then leads to a description of the Cormack-Jolly-Seber (CJS) model. The CJS model, originally conceived at about the same time as the JS model (Cormack 1964), differs from the JS model in that it deals with the marked cohort of animals only. The CJS model conditions on the first capture of each animal and follows the subsequent recapture/reobservation histories throughout the study. Unlike the JS model, the CJS construct does not deal with ratios of marked and unmarked animals and therefore is not able to provide estimates of population size using maximum likelihood theory.

One interesting aspect of the Jolly-Seber model is the assumption that has often been made that the size estimates for open populations are not affected much, if at all, by tag losses or mortality caused by tagging (e.g., Pollock et al. 1990, pp. 25 and 26). This is, in fact, not correct as we have recently demonstrated with both theoretical and simulation results (McDonald et al. 2003).

In chapter 5, James Nichols takes off from the CJS foundation built in Chapter 3 and describes the construction of a likelihood function in terms of the probabilities of survival and capture of the animals in the study. He also describes how those probabilities can be interpreted as functions of the values of covariates that are recorded during the collection of the individual capture histories. The covariates that may help explain capture histories could include the temperatures or other weather parameters at the times of animal captures or in the season prior to the capture events, the characteristics of the researcher's capture efforts (e.g., the number of trap nights, or the numbers of hours flown by helicopter), or the characteristics of the animals themselves (e.g., the sex, age, weight, or reproductive status). The survival and capture probabilities can in this way be treated as if they are constant over the entire sampling period, or they can be assumed to vary with the specified covariates.

The whole approach is very flexible because if available covariates can help explain capture histories, then restrictive assumptions about survival and capture probabilities are not necessary. Also, the maximum likelihood method allows the fit of nested models (models including covariates that are subsets of a more complete cast of covariates) to be compared using chi-squared tests. Further, the appropriateness of any assortment of alternative models can be compared using Akaike's information criterion (AIC) and related methods. For all of these reasons, we recommend the CJS approach, incorporating explanatory covariates, for analyzing capture–recapture data from open populations involving more than three samples.

Chapter 5 also covers some recent extensions to the basic covariate approach to the CJS model. Reverse-time modeling, temporal symmetry models, and the robust design all are theoretically appealing extensions of the basic approach of CJS, and all can employ information from co-variates. Application of these extensions in the real world, however, has been rare due primarily to their recent development.

As noted above, an important limitation with the CJS approach is that it does not directly provide estimates of population sizes. At one time the prevailing opinion among statisticians seemed to be that this was not an important shortcoming. In reality, however, biologists and managers are usually very interested in estimating population sizes and trends. An important innovation therefore is the calculation of Horvitz-Thompson estimates to overcome this limitation (McDonald and Amstrup 2001; see also chapter 9). This method requires only the assumption that the animals captured at any time are representative of the population of animals alive at that time. Because this assumption is fundamental to all capture–recapture methods that draw inferences to the population being studied it is not a serious limitation. With the Horvitz-Thompson method, the estimated population size at the time of a capture occasion is obtained by estimating the capture probability for each of the animals captured at that occasion and adding up the reciprocals of these probabilities. We therefore recommend, the fitting of CJS model for open populations followed by Horvitz-Thompson estimation when population sizes at each occasion are desired.

10.4 Tag-recovery Models

Models for analyzing tag-recovery data were originally developed separately from models for mark–recapture data. The situations of data collection were different from the normal capture and recapture paradigms in that tagged animals usually were recovered dead. Classic tag-recovery data derived from individual animals (most commonly birds or fish) that were tagged and released, and then harvested at a later time. In modeling terms this is not the same as tagging individuals and then recapturing some of them at later times while they are still alive.

In chapter 6 John Hoenig, Kenneth Pollock, and William Hearn describe the analysis of tag-recovery data with the pioneering methods of Brownie et al. (1985). They then explore the situation where an estimate of the exploitation rate (the fraction of animals harvested) explains only a portion of the recovery rate (which also depends on the probability of tag loss, and the probability of a tag on a harvested animal being reported). The chapter concludes with a discussion of methods for testing whether the assumptions of tag-recovery models apply for the data being considered, and the remedies that are available if this is not the case.

The tag-recovery models and methods are particularly appropriate for many fisheries studies where a good estimate of the exploitation rate is crucial for assessing the status of stocks subjected to commercial and or recreational fisheries. Another important application has been waterfowl harvest monitoring. Many species of waterfowl can be marked in great numbers as flightless chicks or during flightless periods that accompany molt. Then, despite extensive migrations, the nationwide network of sport hunters can provide returns in large numbers.

10.5 Other Models

Many wildlife and fish studies have data available from both live recaptures and dead recoveries. For example, waterfowl tagged in the spring are subject to an autumn hunt. Subsequently, birds not harvested may be reobserved on wintering grounds or when they return the next season to the spring nesting and molting areas. To address these cases Richard Barker in chapter 7 describes an extension of the general CJS approach that can incorporate tag returns. He points out that the alternative to modeling the combined data, analyzing each type of data separately, may be considerably less efficient than using the combined approach. Unless it happens that almost all of the data collected in a study are of one type or the other, we recommend the methods described by Barker for studies that produce data from both harvests and live recoveries.

Carl Schwarz's final chapter on methods for analyzing capture–recapture data describes multistrata models, which are more commonly now called multistate models. These models apply where the captures and recaptures made on a population can occur at different geographic locations or "strata," or "states." States may be different geographic locales or behavioral or reproductive conditions (e.g., hibernation, occurrence above or below the surface of the soil, snow, or water, or different reproductive conditions), and where the animals may move between those states from one capture occasion to the next. The CJS model is extended to this situation by allowing capture and survival probabilities to vary with the different states, and by introducing parameters that describe the probabilities of movement among states. Needless to say, this makes the model considerably more complicated than the usual CJS model. Because estimation of the probabilities of movement among states depends on the numbers of observations (captures) in all of them, realistic estimates of all interchange rates among states cannot usually be obtained unless many (often thousands of) animals are tagged, and the numbers of reobservations in each state are very high. With smaller data sets, constraining certain interchange rates to be equal is often necessary, and if done usually results in reliable estimates of the constrained parameters.

In addition to the extension of the CJS model to cases with two or more states, Schwarz also covers closed-population situations, with two samples in particular. He notes, however, that there are several practical problems involved with these two sample situations, including again the need for very large sample sizes.

Conceptually, multistate models seem to fit many biological situations. Models that acknowledge movements of animals among states are intuitively appealing, because we know animals do so. It is theoretically possible to estimate the population sizes in different states at different times for an open population, but because the data sets required to do so are quite large, a more effective approach may be to fit a CJS type of model, extended to take into account the stratification. The population size in each state and at each occasion can then be estimated using the Horvitz-Thompson approach of adding up the reciprocals of capture probabilities for all of the animals captured in that stratum at that time (McDonald and Amstrup 2001). In general, the data requirements and the computational difficulties of multistate modeling of capture–recapture data mean that biologists should think carefully before embarking on studies of this type.

10.6 Model Selection

A few words of caution are in order about automatic methods of model selection such as AIC and its variations. For example, although Manly et al. (1999) found that model selection using this approach usually chose a model that was close to being correct, this was not always the case.

It may be that one or more of a series of plausible models should not be used for estimation, although they may apparently be the best choice using an AIC type of criterion. This was discovered by Boyce et al. (2001) with a simulation study of the estimation of the number of female bears with cubs of the year in Yellowstone National Park. Six plausible models were defined based on the assumption that the numbers of females with cubs observed in any year has a negative binomial distribution. The simulation study demonstrated clearly that the two most complicated models gave such poor estimates of true population sizes that neither of them should ever be used, even when they are known to be the correct model. Similarly, Amstrup et al. (2001) chose not to use the model with the lowest value for an AIC criterion when analyzing mark–recapture data on polar bears because of the poor estimates of population sizes produced by this model.

Our recommendation in terms of model selection by AIC and other similar types of criterion is that these should be regarded as useful guides for choosing models, but that the final decision on a model to use should take into account other factors, such as what management decisions will be based on the model, the apparent quality of estimates of quantities derived indirectly from the estimated parameters in the fitted model, and the degree to which estimates agree with other known information about the populations in question as well as the biologist's intuition regarding the situation, etc. In other words, it is not safe to assume that mathematics alone will always guide you to the correct choice of models.

10.7 Known Ages

We conclude this final chapter with a brief description of a method that has recently been developed for the estimation of open-population sizes from mark–recapture data on animals of known ages (Manly et al. 2003). Age information is used only indirectly, if at all, with the methods described in earlier chapters. For example, in an open-population model survival rates can be assumed to depend on age through the use of a covariate. However, the age information is not used to determine when the animal was in the population, and hence possibly available for capture before the first actual capture time.

There are many cases of long-lived animals where the ages of the individuals can be and routinely are accurately determined. For example, the age of polar bears (*Ursus martimus*) can be determined by removing and sectioning a vestigial premolar tooth. Hence, if a polar bear is captured for the first time and found to be ten years old, then it is known that this bear was in the population and available for capture, provided that the bear did not immigrate into the population after it was born, in each of the previous nine years, even if it was not captured in those years.

This idea can be combined with the Manly and Parr (1968, see chapter 3) concept of estimating the probability of capture by defining a group of animals known to be alive at the time of a sample, and finding what fraction of these animals were actually captured. Age information allows the size of the group to be enlarged by the inclusion of animals captured for the first time in later samples that must have been alive when the sample in question was taken.

A likelihood function can be constructed using age information and conditioning on the last capture time of animals. This turns out to involve capture probability parameters, but not survival probability parameters, so that it is relatively simple in comparison with other likelihood functions for open-population data. It also turns out that the capture probability parameters can easily be interpreted as functions of covariates, in a similar way to the models discussed by James Nichols in chapter 5. There are three considerable advantages of the analysis using age data. First, all the calculations with can be carried out using an ordinary program for logistic regression. The usual specialized software for fitting capture–recapture models is therefore no longer required. This is important because most biologists and managers are familiar with regression concepts and the programs available to them. It is also important because the outputs of such programs are easily interpreted. Second, it is possible to estimate the population size at the time of the first sampling occasion, which is not possible with the Jolly-Seber method, for example. This means that it is possible to estimate population size in only two occasions rather than the minimum of three normally required for open populations. Because funding for long-term studies is always more difficult to obtain, this benefit offers the promise of estimates that otherwise may not have been obtained due to absence of protracted funding commitments. Third, estimates of population size should have better precision than those obtained from other methods that do not use the age information.

A likelihood function can also be constructed conditioning on the known time when captured animals entered the population based on their ages. By treating this time as a first "capture" it becomes possible to model both capture and survival probabilities, using the computer program MARK, for example. This is more complicated than using the logistic regression approach for modeling capture probabilities only, but this may be worthwhile because of the extra information obtained from the times of last captures of animals.

The strengths and weaknesses of this approach to population size estimation are still being developed. Simulation results and recommendations for appropriate application situations should be published in the coming years.`

Figure 10.1. Double ear-tagged black bear (*Ursus americanus*) cub watches mom from an elevated perch, Boise National Forest, Idaho, 1973. (Photo by Steven C. Amstrup)

Appendix

This appendix contains a table listing the capture–recapture software programs and capabilities that the editors either knew about or were able to find on the Internet in August 2002 (Table A1). Also included is a table listing the contact information for each program (Table A2).

TABLE A.1
Capability matrix for common capture–recapture software packages

Program Name	Interactive interface	Text-based interface	Data manipulation	Capture history input	m-Array input	Open-population models	Closed-population models	User-defined models	Unequal time intervals	Parameters fixing	Group covariates	Individual covariates	Abundance estimates	Recruitment estimates	Simulation capability	Variance component estimation	AIC model ranking	Goodness-of-fit testing	Model averaging
					General capabilities														
MARK	+			+	+[a]	+	+	+	+	+	+	+	+	+	+	+	+	+	+
POPAN-5	+	+	+	+		+	+	+	+	+	+		+	+	+		+	+	
SURPH 2.1	+			+		+		+			+	+					+	+	
CARE-2	+			+			+				+	+	+				+		
SURGE 4.2	+			+		+		+		+	+						+		
JOLLY/JOLLYAGE		+		+	+	+			+				+	+					
RELEASE		+			+	+									+			+	
MULT	+					+			+		+							+	
CAPTURE/2 CAPTURE	+			+			+						+		+			+	
ESTIMATE		+			+	+												+	
NOREMARK	+						+						+		+				

Note. +, presence of a capability.
[a] MARK only allows m-array input for dead recovery data.

TABLE A.1 *Continued*

Types of models

Program Name	Cormack-Jolly-Seber	Jolly-Seber	Live recapture	Dead recovery	Brownie et al. dead recovery	Joint encounter	Barker live recapture and dead recovery	Burnham live recapture and dead recovery	Known fate	Robust design	Barker robust design	Robust design with Huggins estimator	Multistate	Closed capture with Hete-rogeneity	Huggins closed capture	Pradel recruitment only	Pradel survival and seniority	Pradel survival and lambda	Nest survival	BTO ring recoveries	Occupancy estimation	Robust design occupancy
MARK	+	+	+	+	+	+	+	+	+	+	+	+	+	+	+	+	+	+	+	+	+	+
POPAN-5	+	+																				
SURPH 2.1	+		+																			
CARE-2			+						+					+	+							
SURGE 4.2	+		+																			
JOLLY/JOLLYAGE		+																				
RELEASE	+		+																			
MULT				+	+																	
CAPTURE/2CAPTURE			+											+								
ESTIMATE				+																		
NOREMARK			+																			

Note. +, presence of a capability.

[a] MARK only allows *m*-array input for dead recovery data.

TABLE A.2
General and contact information for common capture–recapture software packages listed in table A.1

Program	Website	Program contact	Key documentation	Comment
MARK	http://www.cnr .colostate.edu/ ~gwhite/mark/ mark.htm	Gary White, Colorado State University, gwhite@ cnr.colostate.edu	White, G. C., and E. G. Cooch. 2002. Program MARK—A Gentle Introduction . . . , 2nd ed.	MARK is a comprehensive program for capture–recapture analysis. MARK includes the capabilities of most other programs.
POPAN	http://www.cs .umanitoba.ca/ ~popan/	Population Analysis Software Group, University of Manitoba, popan@ cs.umanitoba.ca	Arnason, A. N., C. J. Schwarz, and G. Boyer. 1998. POPAN-5: A data maintenance and analysis system for mark–recapture data, Release 5.0.	POPAN is useful for open population estimates of abundance and recruitment (nonconditional models), stratified, and state–space analyses.
SURPH 2.1	http://www.cbr .washington.edu/ paramEst/SURPH/	Columbia Basin Research, University of Washington, surph@cbr .washington.edu	Lady, J., P. Westhagen, and J. R. Skalski. 2001. SURPH 2.1: SURvival Under Proportional Hazards.	SURPH performs conditional analysis and allows individual covariates.
CARE-2	http://chao.stat.nthu .edu.tw/software CE.html	Anne Chao, National Tsing-Hua University (Taiwan), chao@ stat.nthu.edu.tw	Chao, A. 2002. User Guide for Program CARE-2, Version 1.1.	CARE-2 provides abundance estimates for a variety of closed-population models.

TABLE A.2 *Continued*

Program	Website	Program contact	Key documentation	Comment
SURGE 4.2	http://www.phidot.org/software/surge/	Evan Cooch, Cornell University, evan.cooch@cornell.edu	Cooch, E. G., R. Pradel, and N. Nur. 1996. A Practical Guide to Mark–Recapture Analysis Using SURGE (2nd ed.). Centre d'Ecologie Fonctionelle et Evolutive–CNRS, Montpellier, France, 135 pp.	SURGE performs conditional analysis with individual covariates.
JOLLY and JOLLYAGE	http://www.mbr-pwrc.usgs.gov/software.html	Jim Hines, Patuxent Wildlife Center, jim_hines@usgs.gov	Pollock, K. H., J. D. Nichols, C. Brownie, and J. E. Hines. 1990. Statistical Inference for Capture–Recapture Experiments. Wildlife Monographs 107.	JOLLY and JOLLYAGE provide estimates of abundance and recruitment (nonconditional models). Capabilities have been incorporated into MARK and POPAN.
RELEASE	http://www.cnr.colostate.edu/~gwhite/software.html	Gary White, Colotado State University, gwhite@cnr.colostate.edu	Burnham, K. P., D. R. Anderson, G. C. White, C. Brownie, and K. P. Pollock. 1987. Design and analysis of methods for fish survival experiments based on release-recapture. *Am. Fish. Soc. Monogr.* 5:1–437.	RELEASE can be used for simulation and goodness-of-fit diagnostics. Capabilities have been incorporated into MARK and other software packages.

Program	URL	Contact	Citation	Description
MULT	http://www.mbr-pwrc.usgs.gov/software.html	Michael Conroy, Georgia Cooperative Fish and Wildlife Research Unit, conroy@arches.uga.edu	Conroy, M. J., J. E. Hines, and B. K. Williams. 1989. Procedures for the Analysis of Band-recovery Data and User Instructions for Program MULT. U.S. Fish & Wildlife Service, Resource Publication 175, Washington, DC.	MULT performs analysis of band recovery data, and allows reward-band data.
CAPTURE and 2CAPTURE	http://www.cnr.colostate.edu/~gwhite/software.html	Gary White, Colorado State University, gwhite@cnr.colostate.edu	White, G. C., Anderson, D. R., Burnham, K. P. and D. L. Otis. 1982. Capture-recapture and removal methods for sampling closed populations. Report LA-8787-NERP. Los Alamos National Laboratory, Los Alamos, NM.	CAPTURE and 2CAPTURE perform analyses of closed populations data. Capabilities have been incorporated into MARK.
ESTIMATE	http://www/cnr.colostate.edu/~gwhite/software.html	Gary White, Colorado State University, gwhite@cnr.colostate.edu	Brownie, C., D. R. Anderson, K. P. Burnham, and D. R. Robson. 1985. Statistical Inference from Band Recovery Data—A Handbook, 2nd ed. U.S. Fish Wildl. Serv. Resour. Publ. 131.	ESTIMATE performs conditional analyses. Capabilities have been incorporated into MARK.
NOREMARK	http://www.cnr.colostate.edu/~gwhite/software.html	Gary White, Colorado State University, gwhite@cnr.colostate.edu	White, G. C. 1996. Program NOREMARK Software Reference Manual.	NOREMARK provides abundance estimates from resighting and recapture data for populations with known numbers of marked individuals.

References

Agresti, A. 1994. Simple capture–recapture models permitting unequal catchability and variable sampling effort. *Biometrics* 50:494–500.

Akaike, H. 1973. Information theory and an extension of the maximum likelihood principle. Pages 267–281 *in* B. N. Petran and F. Csàaki, editors. *International Symposium on Information Theory*, 2nd ed. Akadèemiai Kiadi, Budapest, Hungary.

Alexander, H. M., N. A. Slade, and W. D. Kettle. 1997. Application of mark–recapture models to estimation of the population size of plants. *Ecology* 78: 1230–1237.

Alho, J. M. 1990. Logistic regression in capture–recapture models. *Biometrics* 46:623–635.

Amstrup, S. C. 2003. Polar bear, *Ursus maritimus*. Pages 587–610 *in* G. A. Feldhamer, B. C. Thompson, and J. A. Chapman, editors. *Wild mammals of North America: biology, management, and conservation*. John Hopkins University Press, Baltimore, MD.

Amstrup, S. C., T. L. McDonald, and I. Stirling. 2001. Polar bears in the Beaufort Sea: a 30-year mark–recapture case history. *Journal of Agricultural, Biological, and Environmental Statistics* 6(2):221–234.

Andersen, P. K., Ø. Borgan, R. D. Gill, and N. Keiding. 1993. *Statistical models based on counting process*. Springer-Verlag, New York.

Anderson, D. R. 1975. Population ecology of the mallard, V: temporal and geographic estimates of survival, recovery and harvest rates. *U.S. Fish and Wildlife Service, Resource Publication 125*. 110 pp.

Anderson, D. R., and K. P. Burnham. 1976. Population ecology of the mallard, VI: the effect of exploitation on survival. *United States Fish and Wildlife Service, Resource Publication 128*. 66 pp.

Anderson, D. R., and R. T. Sterling. 1974. Population dynamics of molting pintail drakes banded in south-central Saskatchewan. *Journal of Wildlife Management* 38:226–274.

Anderson, D. R., K. P. Burnham, and G. C. White. 1994. AIC model selection in overdispersed capture–recapture data. *Ecology* 75:1780–1793.

Arnason, A. N. 1972. Parameter estimates for mark–recapture experiments on two populations subject to migration and death. *Researches on Population Ecology* 13:99–113.

Arnason, A. N. 1973. The estimation of population size, migration rates, and survival in stratified populations. *Researches on Population Ecology* 15:1–8.

Arnason, A. N, and K. H. Mills. 1981. Bias and loss of precision due to tag loss in Jolly-Seber estimates for mark–recapture experiments. *Canadian Journal of Fisheries and Aquatic Sciences* 38:1077–1095.

Arnason, A. N., and C. J. Schwarz. 1999. Using POPAN-5 to analyse banding data. *Bird Study* 46:S157–S168.

Arnason, A. N., C. W. Kirby, and C. J. Schwarz. 1996a. Stratification can improve the accuracy of mark–recapture estimates of salmon run size. Unpublished technical report.

Arnason, A. N., C. W. Kirby, C. J. Schwarz, and J. R. Irvine. 1996b. Computer analysis of marking data from stratified populations for estimation of salmonid escapements and the size of other populations. *Canadian Technical Report of Fisheries and Aquatic Science 2106.*

Arnason, A. N., C. J. Schwarz, and G. Boyer. 1998. *POPAN-5: a data maintenance and analysis system for mark–recapture data.* Scientific Report, Department of Computer Science, University of Manitoba, Winnipeg. viii+318pp.

Bailey, N.T.J. 1951. On estimating the size of mobile populations from recapture data. *Biometrika* 38:293–306.

Bailey, N.T.J. 1952. Improvements in the interpretation of recapture data. *Journal of Animal Ecology* 21:120–127.

Bailey, N.T.J. 1969. Trap response of wild cottontails. *Journal of Wildlife Management* 33:48–58.

Banneheka, S. G., R. D. Routledge, and C. J. Schwarz. 1997. Stratified two-sample tag–recovery census of closed populations. *Biometrics* 53:1212–1224.

Barker, R. J. 1995. *Open population mark–recapture models including ancillary sightings.* Ph.D. Thesis, Massey University, New Zealand.

Barker, R. J. 1997. Joint modeling of live-recapture, tag–resight, and tag–recovery data. *Biometrics* 53:666–677.

Barker, R. J. 1999. Joint analysis of mark–recapture, resighting and ring-recovery data with age-dependence and marking-effect. *Bird Study* 46:S82–S91.

Barker, R. J., and L. Kavalieris. 2001. Efficiency gain from auxiliary data requiring additional nuisance parameters. *Biometrics.* 57:563–566.

Barker, R. J., K. P. Burnham, and G. C. White. 2004. Encounter history modeling of joint mark–recapture, tag-resighting and tag-recovery data under temporary emigration. *Statistica Sinica.* 14:1037–1055.

Becker, N. G. 1984. Estimating population size from capture–recapture experiments in continuous time. *Australian Journal of Statistics* 26:1–7.

Becker, N. G., and C. C. Heyde. 1990. Estimating population size from multiple recapture experiments. *Stochastic Processes and Their Applications* 36: 77–83.

Bennetts, R. E., J. D. Nichols, J.-D. Lebreton, R. Pradel, J. E. Hines, and W. M. Kitchens. 2001. Methods for estimating dispersal probabilities and related parameters using marked animals. Pages 3–17 *in* J. Clobert, E. Danchin, A. A. Dhondt, and J. D. Nichols, editors. *Dispersal: individual, population, and community.* Oxford University Press, Oxford, UK.

Besbeas, P., S. N. Freeman, B.J.T. Morgan, and E. A. Catchpole. 2002. Integrating mark–recapture–recovery and census data to estimate animal abundance and demographic parameters. *Biometrics* 58:540–547.

Besbeas, P., J.-D. Lebreton, and B.J.T. Morgan. 2003. The efficient integration of abundance and demographic data. *Applied Statistics* 52:95–102.

Borchers, D. L., W. Zucchini, and R. Fewster. 1998. Mark–recapture models for line transect surveys. *Biometrics* 54:1207–1220.

Boyce, M. S., D. I. Mackenzie, B.F.J. Manly, M. A. Haroldson, and D. Moddy. 2001. Negative binomial models for abundance estimation of multiple closed populations. *Journal of Wildlife Management* 65:498–509.

Bradshaw, C. J., R. J. Barker, and L. S. Davis. 2000. Modeling tag loss in New Zealand fur seal pups. *Journal of Agricultural, Biological, and Ecological Statistics* 5:475–485.

Briand L. C., K. El Eman, B. G. Freimut, and O. Laitenberger. 2000. A comprehensive evaluation of capture–recapture models for estimating software defect content. *IEEE Transactions on Software Engineering* 26:518–538.

Brownie, C. 1973. *Stochastic models allowing age-dependent survival rates for banding experiments on exploited bird populations.* Ph.D. Thesis, Cornell University, Ithaca, NY.

Brownie, C., and D. S. Robson. 1983. Estimation of time-specific survival rates from tag-resighting samples: a generalization of the Jolly-Seber model. *Biometrics* 39:437–453.

Brownie, C., D. R. Anderson, K. P. Burnham, and D. S. Robson. 1978. Statistical inference from band recovery data: a handbook. *United States Fish and Wildlife Service, Resource Publication 131*, Washington, DC.

Brownie, C., D. R. Anderson, K. P. Burnham, and D. S. Robson. 1985. Statistical inference from band recovery data: a handbook, 2nd ed. *United States Fish and Wildlife Service, Resource Publication 156*, Washington, DC.

Brownie, C., J. E. Hines, and J. D. Nichols. 1986. Constant-parameter capture–recapture models. *Biometrics* 42:561–574.

Brownie, C., J. E. Hines, J. D. Nichols, K. H. Pollock, and J. B. Hestbeck. 1993. Capture–recapture studies for multiple strata including non-Markovian transition probabilities. *Biometrics* 49:1173–1187.

Buckland, S. T. 1980. A modified analysis of the Jolly-Seber capture–recapture model. *Biometrics* 36:419–435.

Buckland, S. T. 1982. A mark–recapture survival analysis. *Journal of Animal Ecology* 51:833–847.

Buckland, S. T., and P. H. Garthwaite. 1991. Quantifying precision of mark–recapture estimates using the bootstrap and related methods. *Biometrics* 47:255–268.

Buckland, S. T., K. P. Burnham, and N. H. Augustin. 1997. Model selection: an integral part of inference. *Biometrics* 53:603–618.

Buckland, S. T., I.B.J. Goudie, and D. L. Borchers. 2000. Wildlife population assessment: past developments and future directions. *Biometrics* 56:1–12.

Bulmer, M. 1974. On fitting the Poisson lognormal distribution of species-abundance data. *Biometrics* 30:101–110.

Bunge, J., and M. Fitzpatrick. 1993. Estimating the number of species: a review. *Journal of the American Statistical Association* 88:364–373.

Burnham, K. P. 1972. *Estimation of population size in multiple capture–recapture studies when capture probabilities vary among animals.* Thesis, Oregon State University, Corvallis.

Burnham, K. P. 1993. A theory for combined analysis of ring recovery and recapture data. Pages 199–213 *in* J. D. Lebreton and P. M. North, editors. *The study of bird population dynamics using marked individuals.* Birkhauser Verlag, Berlin.

Burnham, K. P., and D. R. Anderson. 1992. Data-based selection of an appropriate biological model: the key to modern data analysis. Pages 16–30 *in* D. R. McCullough, and R. H. Barrett, editors. *Wildlife 2001: populations.* Elsevier Applied Science, New York.

Burnham, K. P., and D. R. Anderson. 1998. *Model selection and inference: a practical information–theoretic approach.* Springer-Verlag, New York.

Burnham, K. P., and D. R. Anderson. 2002. *Model selection and multimodel inference,* 2nd ed. Springer-Verlag, New York.

Burnham, K. P., and W. S. Overton. 1978. Estimation of the size of a closed population when capture probabilities vary among animals. *Biometrika* 65:625–633.

Burnham, K. P., D. R. Anderson, G. C. White, C. Brownie, and K. H. Pollock. 1987. Design and analysis of methods for fish survival experiments based on release–recapture. *American Fisheries Society Monograph 5.* 437pp.

Campbell, R. P., T. J. Cody, C. E. Bryan, G. C. Matlock, M. F. Osborn, and A. W. Green. 1992. An estimate of the tag-reporting rate of commercial shrimpers in two Texas bays. *United States Fishery Bulletin* 90:621–624.

Carothers, A. D. 1973a. Capture–recapture methods applied to a population with known parameters. *Journal of Animal Ecology* 42:125–146.

Carothers, A. D. 1973b. The effects of unequal catchability on Jolly-Seber estimates. *Biometrics* 29:79–100.

Castledine, B. J. 1981. A Bayesian analysis of multiple-recapture sampling for a closed population. *Biometrika* 68:197–210.

Caswell, H. 2001. *Matrix population models.* Sinauer, Sunderland, MA.

Catchpole, E. A., and B. J. T. Morgan. 1997. Detecting parameter redundancy. *Biometrika* 84:197–196.

Catchpole, E. A., S. N. Freeman, B. J. T. Morgan, and M. P. Harris. 1998. Integrated recovery/recapture data analysis. *Biometrics* 54:33–46.

Chao, A. 1987. Estimating the population size for capture–recapture data with unequal catchability. *Biometrics* 43:783–791.

Chao, A. 1988. Estimating animal abundance with capture frequency data. *Journal of Wildlife Management* 52:295–300.

Chao, A. 1989. Estimating population size for sparse data in capture–recapture experiments. *Biometrics* 45:427–438.

Chao, A. 2001. An overview of closed capture–recapture models. *Journal of Agricultural, Biological and Environmental Statistics* 6:158–175.

Chao, A. 2005. Species richness estimation. *In* N. Balakrishnan, C. B. Read and B. Vidakovic, editors. *Encyclopedia of statistical sciences,* 2nd ed. Wiley, New York. (In press.)

Chao, A., and Bunge, J. 2002. Estimating the number of species in a stochastic abundance model. *Biometrics* 58:531–539.

Chao, A., and S.-M. Lee. 1992. Estimating the number of classes via sample coverage. *Journal of the American Statistical Association* 417:210–217.

Chao, A., and S.-M. Lee. 1993. Estimating population size for continuous-time capture–recapture models via sample coverage. *Biometrical Journal* 35:29–45.

Chao, A., S.-M. Lee, and S.-L. Jeng. 1992. Estimating population size for capture–recapture data when capture probabilities vary by time and individual animal. *Biometrics* 48:201–216.

Chao, A., W. Chu, and C.-H. Hsu. 2000. Capture–recapture when time and behavioral response affect capture probabilities. *Biometrics* 56:427–433.

Chao, A., P. K. Tsay, S.-H. Lin, W.-Y. Shau, and D.-Y. Chao. 2001a. Tutorial in biostatistics: the applications of capture–recapture models to epidemiological data. *Statistics in Medicine* 20:3123–3157.

Chao, A., P.S.F. Yip, S.-M. Lee, and W. Chu. 2001b. Population size estimation based on estimating functions for closed capture–recapture models. *Journal of Statistical Planning and Inference* 92:213–232.

Chapman, D. H. 1951. Some properties of the hypergeometric distribution with applications to zoological censuses. *University of California Publications in Statistics* 1:131–160.

Chapman, D. H., and C. O. Junge. 1956. The estimation of the size of a stratified animal population. *Annals of Mathematical Statistics* 27:375–389.

Choquet, R., A. M. Reboulet, R. Pradel, O. Gimenez, and J.-D. Lebreton. 2003. *User's Manual for M-SURGE 1.01*. CEFE/CNRS, Montpellier (http://www .cefe.cnrs-mop.fr/wwwbiom/Soft-CR/m-surge.htm).

Clobert, J., and J.-D. Lebreton. 1985. Dependance de facteurs de milieu dans les estimations de taux de survie par capture–recapture. *Biometrics* 41:1031–1037.

Clobert, J., J.-D. Lebreton, M. Clobert-Gillet, and H. Coquillart. 1985. The estimation of survival in bird populations by recaptures or resightings of marked individuals. Pages 197–213 *in* B.J.T. Morgan and P. M. North, editors. *Statistics in Ornithology*. Springer-Verlag, New York.

Clobert, J., J.-D. Lebreton, and D. Allaine. 1987. A general approach to survival rate estimation by recaptures or resightings of marked birds. *Ardea* 75:133–142.

Clobert, J., J.-D. Lebreton, and G. Marzolin. 1990. The estimation of local immature survival rates and of age-specific proportions of breeders in bird populations. Pages 199–213 *in* J. Blondel, A. Gosler, J.-D. Lebreton, and R. H. McCleery, editors. *Population biology of passerine birds: an integrated approach*. Springer-Verlag, Berlin.

Clobert, J., J.-D. Lebreton, D. Allaine, and J.-M. Gaillard. 1994. The estimation of age-specific breeding probabilities from recaptures or resightings in vertebrate populations, II: longitudinal models. *Biometrics* 50:375–387.

Cochran, W. G. 1978. Laplace's ratio estimators. Pages 3–10 *in* H. A. David, editor. *Contributions to survey sampling and applied statistics*. Academic Press, New York.

Colwell, R. K. 1997. *Estimates: statistical estimation of species richness and shared species from samples*, Version 5. User's guide and application published at http://viceroy.eeb.uconn.edu/estimates.

Conroy, M. J., and B. K. Williams. 1984. A general methodology for maximum likelihood inference from band recovery data. *Biometrics*. 40: 739–748.

Cooch, E. G., and G. White. 2001. *Using MARK: a gentle introduction*, 2nd ed. Available on line at http://www.phidot.org/software/mark/docs/book/ (July 2002).

Cooch, E. G., R. Pradel, and N. Nur. 1996. *A practical guide to mark–recapture analysis using SURGE*, 2nd ed. Centre d'Ecologie Fonctionelle et Evolutive–CNRS, Montpellier, France. http://www.phidot.org/software/surge/ (June 2002).

Cormack, R. M. 1964. Estimates of survival from the sightings of marked animals. *Biometrika* 51:429–438.

Cormack, R. M. 1968. The statistics of capture–recapture methods. *Oceanography and Marine Biology: An Annual Review* 6:455–506.

Cormack, R. M. 1981. Loglinear models for capture–recapture experiments on open populations. Pages 217–235 *in* R. W. Hirons and D. Cooke, editors. *The mathematical theory of the dynamics of biological populations*, II. Academic Press, London.

Cormack, R. M. 1989. Log–linear models for capture–recapture. *Biometrics* 45: 395–413.

Cormack, R. M. 1992. Interval estimation for mark-recapture studies of closed populations. *Biometrics* 48:567–576.

Costello, T. J., and D. M. Allen. 1968. Mortality rates in populations of pink shrimp, *Penaeus duorarum*, on the Sambel and Tortugas grounds, Florida. *United States Fishery Bulletin* 66:491–502.

Coull, B. A., and A. Agresti. 1999. The use of mixed logit models to reflect heterogeneity in capture–recapture studies. *Biometrics* 55:294–301.

Craig, C. C. 1953. On the utilization of marked specimens in estimating populations of flying insects. *Biometrika* 40:170–176.

Crosbie, S. F., and B.F.J. Manly. 1985. Parsimonious modelling of capture–mark–recapture studies. *Biometrics* 41:385–398.

Crowder, M. J., A. C. Kimber, R. L. Smith, and T. J. Sweeting. 1991. *Statistical analysis of reliability data*. Chapman & Hall, London.

Darroch, J. N. 1958. The multiple-recapture census, I: estimation of a closed population. *Biometrika* 45:343–359.

Darroch, J. N. 1959. The multiple-recapture census, II: estimation when there is immigration or death. *Biometrika* 46:336–351.

Darroch, J. N. 1961. The two-sample capture–recapture census when tagging and sampling are stratified. *Biometrika* 48:241–260.

Darroch, J. N., S. E. Feinberg, G. F. V. Glonek, and B. W. Junker. 1993. A three-sample multiple-recapture approach to census population estimation with heterogeneous catchability. *Journal of the American Statistical Association* 88: 1137–1148.

DeMaster, D. P., and I. Stirling. 1981. *Ursus maritimus*: Polar bear. *Mammalian Species* 145:1–7.

Diefenbach, D. R., and G. L. Alt. 1998. Modeling and evaluation of ear tag loss in black bears. *Journal of Wildlife Management* 62:1292–1300.

Dupuis, J. A. 1995. Bayesian estimation of movement and survival probabilities from capture–recapture data. *Biometrika* 82:761–772.

Dupuis, J. A. 2002. Prior distributions for stratififed capture–recapture models. *In* B. J. T. Morgan and D. L. Thomson, editors. Statistical analysis of data from marked bird populations. *Journal of Applied Statistics* 29:225–237.

Dupuis, J. A. and C. J. Schwarz. 2005. Estimates of abundance, survival, recruitment, and movement using a stratified Jolly-Seber model. In preparation.

Edwards, W. R., and L. L. Eberhardt. 1967. Estimating cottontail abundance from live-trapping data. *Journal of Wildlife Management* 31:87–96.

Efron, B., and R. Thisted. 1976. Estimating the number of unseen species: how many words did Shakespeare know? *Biometrika* 63:435–447.

Efron, B., and R. Tibshirani. 1993. *An introduction to the Bootstrap*. Chapman Hall, New York.

Evans, M. A., D. G. Bonett, and L. L. McDonald. 1994. A general theory for modeling capture–recapture data from a closed population. *Biometrics* 50: 396–405.

Fabrizio, M. C., J. D. Nichols, J. E. Hines, B. L. Swanson, and S. T. Schram. 1999. Modeling data from double-tagging experiments to estimate heterogeneous rates of tag-shedding in lake trout (*Salvelinus namaycush*). *Canadian Journal of Fisheries and Aquatic Sciences* 56:1409–1419.

Fienberg, S. E. 1972. The multiple recapture census for closed populations and incomplete 2^k contingency tables. *Biometrika* 59:591–603.

Fisher, R. A., and E. B. Ford. 1947. The spread of a gene in natural conditions in a colony of the moth *Panaxia dominula*. *Heredity* 1:143–174.

Fisher, R. A., A. S. Corbet, and C. B. Williams. 1943. The relation between the number of species and the number of individuals in a random sample of an animal population. *Journal of Animal Ecology* 12:42–58.

Freeman, S. N., B.J.T. Morgan, and E. A. Catchpole. 1992. On the augmentation of ring recovery data with field information. *Journal of Animal Ecology* 61:649–657.

Frusher, S. D., and J. M. Hoenig. 2001. Estimating natural and fishing mortality and tag reporting rate of rock lobster from a multi-year tagging model. *Canadian Journal of Fisheries and Aquatic Sciences* 58:2490–2501.

Fuller, M. 1981. *Characteristics of an American alligator (Alligator mississippiensis) population in the vicinity of Lake Ellis Simon*. M.S. Thesis, North Carolina State University, Raleigh. 110pp.

Ganter, B. 1995. Site tenacity and movements of staging barnacle geese (*Branta leucopsis*). *Ardea* 82:231–240.

George, E. I., and C. P. Robert. 1992. Capture–recapture estimation via Gibbs sampling. *Biometrika* 79:677–683.

Gimenez, O., R. Choquet, and J.-D. Lebreton. 2003. Parameter redundancy in multistate capture–recapture models. *Biometrical Journal* 45:704–722.

Gilbert, R. O. 1973. Approximations of the bias in the Jolly-Seber capture–recapture model. *Biometrics* 29:501–526.

Green, A. W., G. C. Matlock, and J. E. Weaver. 1983. A method for directly estimating the tag-reporting rate of anglers. *Transactions of the American Fisheries Society* 112:412–415.

Greenwood, M., and G. U. Yule. 1920. An inquiry into the nature of frequency distributions representative of multiple happenings with particular reference to the occurrence of multiple attacks of disease or of repeated accidents. *Journal of the Royal Statistical Society* 83:255–279.

Greenwood, R. J., A. B. Sargeant, and D. H. Johnson. 1985. Evaluation of mark–recapture for estimating striped skunk abundance. *Journal of Wildlife Management* 49:332–340.

Hald, A. 1990. *A history of probability and statistics and their applications before 1750.* Wiley, New York.

Hastings, K. K., and J. W. Testa. 1998. Maternal and birth colony effects on survival of Weddell seal offspring from McMurdo Sound, Antarctica. *Journal of Animal Ecology* 67:722–740.

Hearn, W. S., K. H. Pollock, and E. Brooks. 1998. Pre- and post-season tagging models: estimation of reporting rate and fishing and natural mortality rates. *Canadian Journal of Fisheries and Aquatic Sciences* 55:199–205.

Hearn, W. S., T. Polacheck, K. H. Pollock, and W. Whitelaw. 1999. Estimation of tag reporting rates in age-structured multicomponent fisheries where one component has observers. *Canadian Journal of Fisheries and Aquatic Sciences* 56:1255–1265.

Hearn, W. S., J. M. Hoenig, K. H. Pollock, and D. Hepworth. 2003. Tag reporting rate estimation, 3: use of planted tags in one component of a multicomponent fishery. *North American Journal of Fisheries Management* 23:66–77.

Henny, C. J., and K. P. Burnham. 1976. A reward band study of mallards to estimate band reporting rates. *Journal of Wildlife Management* 40:1–14.

Hestbeck, J. B., J. D. Nichols, and R. A. Malecki. 1991. Estimates of movement and site fidelity using mark–resight data of wintering Canada geese. *Ecology* 72:523–533.

Hines, J. E., and J. D. Nichols. 2002. Investigations of potential bias in the estimation of lambda using Pradel's (1996) model for capture–recapture data. *Journal of Applied Statistics* 29:573–587.

Hoenig, J. M., N. J. Barrowman, W. S. Hearn, and K. H. Pollock. 1998a. Multiyear tagging studies incorporating fishing effort data. *Canadian Journal of Fisheries and Aquatic Sciences* 55:1466–1476.

Hoenig, J. M., N. J. Barrowman, K. H. Pollock, E. N. Brooks, W. S. Hearn, and T. Polacheck. 1998b. Models for tagging data that allow for incomplete mixing of newly tagged animals. *Canadian Journal of Fisheries and Aquatic Sciences* 55:1477–1483.

Hoffman, A., and J. R. Skalski. 1995. Inferential properties of an individual-based survival model using release-recapture data: sample size, validity and power. *Journal of Applied Statistics* 22:579–595.

Horvitz, D. G., and D. J. Thompson. 1952. A generalization of sampling without replacement from a finite universe. *Journal of the American Statistical Association* 47: 663–685.

Huang, S.-P., and B. S. Weir. 2001. Estimating the total number of alleles using a sample coverage method. *Genetics* 159:1365–1373.

Huggins, R. M. 1989. On the statistical analysis of capture experiments. *Biometrika* 76:133–140.

Huggins, R. M. 1991. Some practical aspects of a conditional likelihood approach to capture experiments. *Biometrics* 47:725–732.

Huggins, R. M. 2002. A parametric empirical Bayes approach to the analysis of capture–recapture experiments. *Australian and New Zealand Journal of Statistics* 44:55–62.

Hwang, W.-H., and A. Chao. 2002. Continuous-time capture–recapture models with covariate. *Statistica Sinica* 12:1115–1131.

Hwang, W.-H., A. Chao, and P. S. F. Yip. 2002. Continuous-time capture–recapture models with time variation and behavioral response. *Australian and New Zealand Journal of Statistics* 44:41–54.

International Working Group for Disease Monitoring and Forecasting (IWDGMF). 1995. Capture–recapture and multiple-record systems estimation, I: history and theoretical development. *American Journal of Epidemiology* 142:1047–1058.

Jackson, C.H.N. 1939. The analysis of an animal population. *Journal of Animal Ecology* 8:238–246.

Jackson, C.H.N. 1940. The analysis of a tsetse fly population, I. *Annals of Eugenics* 10:332–369.

Jackson, C. H. N. 1944. The analysis of a tsetse fly population, II. *Annals of Eugenics* 12:176–205.

Jackson, C.H.N. 1948. The analysis of a tsetse fly population, III. *Annals of Eugenics* 14:91–108.

Jolly, G. M. 1965. Explicit estimates from capture–recapture data with both death and immigration: stochastic model. *Biometrika* 52:225–247.

Jolly, G. M. 1982. Mark–recapture models with parameters constant in time. *Biometrics* 38:301–321.

Karanth, K. U., and J. D. Nichols. 1998. Estimation of tiger densities in India using photographic captures and recaptures. *Ecology* 79:2852–2862.

Keating, K. A., C. C. Schwartz, M. A. Haroldson, and D. Moody. 2002. Estimating numbers of females with cubs-of-the-year in the Yellowstone grizzly bear population. *Ursus* 13:161–174.

Kendall, W. L., and R. Bjorkland. 2001. Using open robust design models to estimate temporary emigration from capture–recapture data. *Biometrics* 57:1113–1122.

Kendall, W. L., and J. E. Hines. 1999. Program RDSURVIV: an estimation tool for capture–recapture data collected under Pollock's robust design. *Bird Study* 46:S32–S38.

Kendall, W. L., and J. D. Nichols. 1995. On the use of secondary capture–recapture samples to estimate temporary emigration and breeding proportions. *Journal of Applied Statistics* 22:751–762.

Kendall, W. L. and J. D. Nichols. 2002. Estimating state-transition probabilities for unobservable states using capture–recapture/resighting data. *Ecology* 83:3276–3284.

Kendall, W. L., K. H. Pollock, and C. Brownie. 1995. A likelihood-based approach to capture–recapture estimation of demographic parameters under the robust design. *Biometrics* 51:293–308.

Kendall, W. L., J. D. Nichols, and J. E. Hines. 1997. Estimating temporary emigration using capture–recapture data with Pollock's robust design. *Ecology* 78:563–578.

Kendall, W. L., J. E. Hines, and J. D. Nichols. 2003. Adjusting multistate capture–recapture models for misclassification bias: manatee breeding proportions. *Ecology* 84:1058–1066.

Kimura, D. K. 1976. Estimating the total number of marked fish present in a catch. *Transactions of the American Fisheries Society* 105:664–668.

Krebs, C. J. 1999. *Ecological methodology*, 2nd ed. Benjamin/Cummings, Menlo Park, CA.

Latour, R. J., J. M. Hoenig, J. E. Olney, and K. H. Pollock. 2001a. A simple test for nonmixing in multi-year tagging studies: application to striped bass (*Morone saxatilis*) tagged in the Rappahannock River, Virginia. *Transactions of the American Fisheries Society* 130:848–856.

Latour, R. J., J. M. Hoenig, J. E. Olney, and K. H. Pollock. 2001b. Diagnostics for multi-year tagging models with application to Atlantic striped bass (*Morone saxatilis*). *Canadian Journal of Fisheries and Aquatic Sciences* 58:1716–1726.

Latour, R. J., K. H. Pollock, C. A. Wenner, and J. M. Hoenig. 2001c. Estimates of fishing and natural mortality for subadult red drum (*Sciaenops ocellatus*) in South Carolina waters. *North American Journal of Fisheries Management* 21:733–744.

Lawless, J. F., and K. Thiagarajah. 1996. A point-process model incorporating renewals and time trends, with application to repairable systems. *Technometrics* 38:131–138.

Lebreton, J.-D., and R. Pradel. 2002. Multi-stratum recapture models: modelling incomplete individual histories. *Journal of Applied Statistics* 29:353–369.

Lebreton, J.-D., K. P. Burnham, J. Clobert, and D. R. Anderson. 1992. Modelling survival and testing biological hypotheses using marked animals: a unified approach with case studies. *Ecological Monographs* 62:67–118.

Lebreton, J.-D., T. Almeras, and R. Pradel. 1999. Competing events, mixture of information, and multi-strata recapture models. *Bird Study* 46:S39–S46.

Lebreton, J.-D., J. E. Hines, R. Pradel, J. D. Nichols, and J. A. Spendelow. 2003. Estimation by capture–recapture of recruitment and dispersal over several sites. *Oikos* 101:253–264.

Le Cren, E. D. 1965. A note on the history of mark–recapture population estimates. *Journal of Animal Ecology* 34:453–454.

Lee, S.-M. 1996. Estimating population size for capture–recapture data when capture probabilities vary by time, behavior and individual animal. *Communications in Statistics B—Simulation and Computation* 25:431–457.

Lee, S.-M., and A. Chao. 1994. Estimating population size via sample coverage for closed capture–recapture models. *Biometrics* 50:88–97.

Lee, S.-M., and C. W. S. Chen. 1998. Bayesian inference of population size for behavioral response models. *Statistica Sinica* 8:1233–1247.

Lefebrve, L. W., D. L. Otis, and N. R. Holler. 1982. Comparison of open and closed models for cotton rat population estimates. *Journal of Wildlife Management.* 46:156–163.

Leslie, P. H. 1958. A capture–recapture analysis of a shearwater population: with a statistical appendix. *Journal of Animal Ecology* 27:84–86.

Leslie, P. H., and D. Chitty. 1951. The estimation of population parameters from data obtained by means of the capture recapture method, I: the maximum likelihood equations for estimating the death rate. *Biometrika* 38:269–292.

Leslie, P. H., D. Chitty, and H. Chitty. 1953. The estimation of population parameters from data obtained by means of the capture recapture method, III: an example of the practical applications of the method. *Biometrika* 40:137–169.

Lin, D. Y., and P. S. F. Yip. 1999. Parametric regression models for continuous time removal and recapture studies. *Journal of the Royal Statistical Society* B 61:401–411.

Lindberg, M. S., W. L. Kendall, J. E. Hines, and M. G. Anderson. 2001. Combining band recovery data and Pollock's robust design to model temporary and permanent emigration. *Biometrics* 57:273–281.

Link, W. A. 2003. Nonidentifiability of population size from capture–recapture data with heterogeneous detection probabilities. *Biometrics.* 59:1123–1130.

Lloyd, C. J. 1994. Efficiency of martingale methods in recapture studies. *Biometrika* 81:305–315.

Lloyd, C. J., and P. S. F. Yip. 1991. A unification of inference for capture–recapture studies through martingale estimating functions. Pages 65–88 *in* V. P. Godambe, editor. *Estimating equations.* Clarendon Press, Oxford, UK.

Loery, G., K. H. Pollock, J. D. Nichols, and J. E. Hines. 1987. Age-specificity of black-capped chickadee survival rates: analysis of capture–recapture data. *Ecology* 68:1038–1044.

Lunn, N. J., I. Stirling, D. Andriashek, and G. B. Kolenosky. 1997. Re-estimating the size of the polar bear population in Western Hudson Bay. *Arctic* 50: 234–240.

Macdonald, P. D. M., and H. D. Smith. 1980. Mark–recapture estimation of salmon smolt runs. *Biometrics* 36:401–417.

Mace, R. D., S. C. Minta, T. L. Manley, and K. E. Aune. 1994. Estimating grizzly bear population size using camera sightings. *Wildlife Society Bulletin* 22:74–83.

Manly, B.F.J., and M. J. Parr. 1968. A new method of estimating population size, survivorship and birth rate from capture–recapture data. *Transactions of the Society for British Entomology* 18:81–89.

Manly, B.F.J., L. L. McDonald, and T. L. McDonald. 1999. The robustness of mark–recapture methods: a case study for the northern spotted owl. *Journal of Agricultural, Biological, and Environmental Statistics* 4:78–101.

Manly, B.F.J., T. L. McDonald, S. C. Amstrup, and E. V. Regehr. 2003. Improving size estimates of open animal populations by incorporating information on age. *BioScience* 53:666–669.

Mardekian, S. Z., and L. L. McDonald. 1981. Simultaneous analysis of band-recovery and live-recapture data. *Journal of Wildlife Management* 45:484–488.

Mares, M. A., K. E. Streilein, and M. R. Willig. 1981. Experimental assessment of several population estimation techniques on an introduced population of eastern chipmunks. *Journal of Mammalogy* 62:315–328.

Marzolin, G. 1988. Polyginie du Cincle plongeur (*Cinclus cinclus*) dans les cotes de Lorraine. *L'oiseau et la revue Francaise d' ornithologie* 58:277–286.

McCullagh, P., and J. A. Nelder. 1989. *Generalized linear models.* Chapman & Hall, New York.

McDonald, T. L., and S. C. Amstrup. 2001. Estimation of population size using open capture–recapture models. *Journal of Agricultural, Biological, and Environmental Statistics* 6:206–220.

McDonald, T. L., S. C. Amstrup, and B.F.J. Manly. 2003. Tag loss can bias Jolly-Seber capture–recapture estimates. *Wildlife Society Bulletin* 31:814–822.

Moran, P.A.P. 1951. A mathematical theory of animal trapping. *Biometrika* 38:307–311.

Myers, R. A., and J. M. Hoenig. 1997. Direct estimates of gear selectivity from multiple tagging experiments. *Canadian Journal of Fisheries and Aquatic Sciences* 54:1–9.

Nichols, J. D., and C. J. Coffman. 1999. Demographic parameter estimation for experimental landscape studies of small mammal populations. Pages 287–309 *in* G. W. Barrett and J. D. Peles, editors. *Landscape ecology of small mammals.* Springer-Verlag, New York.

Nichols, J. D., and J. E. Hines. 2002. Approaches for the direct estimation of lambda, and demographic contributions to lambda, using capture–recapture data. *Journal of Applied Statistics* 29:539–568.

Nichols, J. D., and A. Kaiser. 1999. Quantitative studies of bird movement: a methodological review. *Bird Study* 46:S289–S298.

Nichols, J. D., and K. H. Pollock. 1990. Estimation of recruitment from immigration versus in situ reproduction using Pollock's robust design. *Ecology* 71:21–26.

Nichols, J. D., R. S. Pospahala, and J. E. Hines. 1982a. Breeding-ground habitat conditions and the survival of mallards. *Journal of Wildlife Management* 45:80–87.

Nichols , J. D., M. J. Conroy, D. R. Anderson, and K. P. Burnham. 1984a. Compensatory mortality in waterfowl populations: a review of the evidence and implications for research and management. *Transactions North American Natural Resources Conference* 49:535–554.

Nichols, J. D., K. H. Pollock, and J. E. Hines. 1984b. The use of a robust capture–recapture design in small mammal population studies: a field example with *Microtus pennsylvanicus. Acta Theriologica.* 29:357–365.

Nichols, J. D., R. W. Morris, C. Brownie, and K. H. Pollock. 1986. Sources of variation in extinction rates, turnover and diversity of marine invertebrate families during the Paleozoic. *Paleobiology* 12:421–432.

Nichols, J. D., J. A. Spendelow, and J. E. Hines. 1990. Capture–recapture estimation of prebreeding survival rate for birds exhibiting delayed maturation. *Journal of Field Ornithology* 61:347–354.

Nichols, J. D., R. J. Blohm, R. E. Reynolds, R. E. Trost, J. E. Hines, and J. P. Bladen. 1991. Band reporting rates for mallards with reward bands of different dollar values. *Journal of Wildlife Management* 55:119–126.

Nichols, J. D., J. Bart, R. J. Limpert, W. L. Sladen, and J. E. Hines. 1992a. Annual survival rates of adult and immature eastern population tundra swans. *Journal of Wildlife Management* 56:485–494.

Nichols, J. D., C. Brownie, J. E. Hines, K. H. Pollock, J. B. Hestbeck. 1993. The estimation of exchanges among population or sub-populations. Pages 265–279 *in* J.-D. Lebreton and P.M. North, editors. *Marked individuals in the study of bird populations.* Birkhäuser Verlag, Basel.

Nichols, J. D., J. E. Hines, K. H. Pollock, R. L. Hinz, and W. A. Link. 1994. Estimating breeding proportions and testing hypotheses about costs of reproduction with capture–recapture data. *Ecology* 75:2052–2065.

Nichols, J. D., J. E. Hines, and P. Blums. 1997. Tests for senescent decline in annual survival probabilities of common pochards, *Athya ferina. Ecology* 78: 1009–1018.

Nichols, J. D., T. Boulinier, J. E. Hines, K. H. Pollock, and J. R. Sauer. 1998. Estimating rates of local extinction, colonization and turnover in animal communities. *Ecological Applications* 8:1213–1225.

Nichols, J. D., J. E. Hines, J.-D. Lebreton, and R. Pradel. 2000. The relative contributions of demographic components to population growth: a direct estimation approach based on reverse-time capture–recapture. *Ecology* 81:3362–3376.

Nichols, J. D., J. R. Sauer, K. H. Pollock, and J. B. Hestbeck. 1992b. Estimating transition probabilities for stage-based population projection matrices using capture-recapture data. *Ecology* 73:306–312.

Nichols, J. D., S. L. Stokes, J. E. Hines, and M. J. Conroy. 1982b. Additional comments on the assumption of homogeneous survival rates in modern bird banding estimation models. *Journal of Wildlife Management* 46:953–962.

Norris, J. L., and K. H. Pollock. 1995. Capture–recapture model with heterogeneity and behavioural response. *Environmental and Ecological Statistics* 2:305–313.

Norris, J. L., and K. H. Pollock. 1996a. Including model uncertainty in estimating variances in multiple capture studies. *Environmental and Ecological Statistics* 3:235–244.

Norris, J. L., and K. H. Pollock. 1996b. Nonparametric MLE under two closed capture–recapture models with heterogeneity. *Biometrics* 52:639–649.

North, P. M., and B.J.T. Morgan. 1979. Modeling heron survival using weather data. *Biometrics* 35:667–681.

Ord, J. K., and G. A. Whitmore. 1986. The Poisson–inverse Gaussian distribution as a model for species abundance. *Communications in Statistics A—Theory and Methods* 15:853–871.

Otis, D. L., K. P. Burnham, G. C. White, and D. R. Anderson. 1978. Statistical inference from capture data on closed animal populations. *Wildlife Monographs* 62:1–135.

Paulik, G. J. 1961. Detection of incomplete reporting of tags. *Journal of the Fisheries Research Board of Canada* 18:817–832.

Perrins, C. M. 1963. Survival in the Great Tit, *Parus major*. *Proceedings of the International Ornithology Congress* 13:717–728.

Perrins, C. M. 1965. Population fluctuations and clutch size in the great tit, *Parus major L. Journal of Animal Ecology* 34:601–647.

Plante, N., L.-P. Rivest, and G. Tremblay. 1998. Stratified capture–recapture estimation of the size of a closed population. *Biometrics* 54:47–60.

Pledger, S. 2000. Unified maximum likelihood estimates for closed capture-recapture models using mixtures. *Biometrics* 56:434–442.

Pollock, K. H. 1974. *The assumption of equal catchability of animals in tag-recapture experiments*. Thesis, Cornell University, Ithaca, NY.

Pollock, K. H. 1975. A K-sample tag–recapture model allowing for unequal survival and catchability. *Biometrika* 62:577–583.

Pollock, K. H. 1981a. Capture–recapture models allowing for age-dependent survival and capture rates. *Biometrics* 37:521–529.

Pollock, K. H. 1981b. Capture–recapture models: a review of current methods, assumptions, and experimental design. Pages 426–435 *in* C. J. Ralph and J. M. Scott, editors. *Studies in avian biology, 6: Estimating numbers of terrestrial birds*. Allen Press, Lawrence, KS.

Pollock, K. H. 1982. A capture–recapture design robust to unequal probability of capture. *Journal of Wildlife Management* 46:757–760.

Pollock, K. H. 1991. Modeling capture, recapture, and removal statistics for estimation of demographic parameters for fish and wildlife populations: past, present, and future. *Journal of the American Statistical Association* 86: 225–238.

Pollock, K. H. 2000. Capture–recapture models. *Journal of the American Statistical Association* 95:293–296.

Pollock, K. H. 2001. George Seber: a statistical ecology pioneer and scientist par excellence. *Journal of Agricultural, Biological, and Environmental Statistics* 6:152–157.

Pollock, K. H., and R. H. K. Mann. 1983. The use of an age-dependent mark–recapture model in fisheries research. *Canadian Journal of Fisheries and Aquatic Sciences* 40:1449–1455.

Pollock, K. H., and M. C. Otto. 1983. Robust estimation of population size in closed animal populations from capture–recapture experiments. *Biometrics* 39:1035–1049.

Pollock, K. H., and D. G. Raveling. 1982. Assumptions of modern band-recovery models with emphasis on heterogeneous survival rates. *Journal of Wildlife Management* 46:88–98.

Pollock, K. H., D. L. Solomon, and D. S. Robson. 1974. Tests for mortality and recruitment in a K-sample tag–recapture experiment. *Biometrics* 30:77–87.

Pollock, K. H., J. E. Hines, and J. D. Nichols. 1984. The use of auxiliary variables in capture–recapture and removal experiments. *Biometrics* 40: 329–340.

Pollock, K. H., J. E. Hines, and J. D. Nichols. 1985. Goodness-of-fit tests for open capture–recapture models. *Biometrics* 41:399–410.

Pollock, K. H., J. D. Nichols, C. Brownie, and J. E. Hines. 1990. Statistical inference for capture–recapture experiments. *Wildlife Monographs* 107:1–97.

Pollock, K. H., J. M. Hoenig, and C. M. Jones. 1991. Estimation of fishing and natural mortality rates when a tagging study is combined with a creel survey or port sampling. American Fisheries Society Symposium 12:423–434.

Pollock, K. H., J. M. Hoenig, W. S. Hearn, and B. Calingaert. 2001. Tag reporting rate estimation, 1: an evaluation of the reward tagging method. *North American Journal of Fisheries Management* 21:521–532.

Pollock, K. H., J. M. Hoenig, W. S. Hearn, and B. Calingaert. 2002a. Tag reporting rate estimation, II: use of high-reward tagging and observers in multicomponent fisheries. *North American Journal of Fisheries Management* 22: 727–736.

Pollock, K. H., W. S. Hearn, and T. Polacheck. 2002b. A general model for tagging on multi-component fisheries: an integration of age-dependent reporting rates and mortality estimation. *Journal of Environmental and Ecological Statistics* 9:57–69.

Powell, L. A., M. J. Conroy, J. E. Hines, J. D. Nichols, and D. G. Krementz. 2000. Simultaneous use of mark–recapture and radio telemetry to estimate survival, movement, and capture rates. *Journal of Wildlife Management* 64: 302–313.

Pradel, R. 1993. Flexibility in survival analysis from recapture data: handling trap-dependence. Pages 29-37 *in* J.-D. Lebreton and P. M. North, editors. *Marked individuals in the study of bird population.* Birkhauser Verlag, Basel, Switzerland.

Pradel, R. 1996. Utilization of capture–mark–recapture for the study of recruitment and population growth rate. *Biometrics* 52:703–709.

Pradel, R., and J.-D. Lebreton. 1999. Comparison of different approaches to the study of local recruitment. *Bird Study* 46 (Suppl): S74–S81.

Pradel, R., J. Clobert, and J.-D. Lebreton. 1990. Recent developments for the analysis of capture–recapture data sets: an example concerning two blue tit populations. *The Ring* 13:193–204.

Pradel, R., J. E. Hines, J.-D. Lebreton, and J. D. Nichols. 1997a. Capture–recapture survival models taking account of transients. *Biometrics* 53:60–72.

Pradel, R., A. R. Johnson, A. Viallefont, R. G. Nager, and F. Cezilly. 1997b. Local recruitment in the Greater Flamingo: a new approach using capture–recapture data. *Ecology* 78:1431–1445.

Pradel, R., C. M. A. Wintrebert, and O. Gimenez. 2003. A proposal for a goodness-of-fit test to the Arnason-Schwarz multisite-capture–recapture model. *Biometrics* 59:43–53.

Pugesek, B. H., C. Nations, K. L. Diem, and R. Pradel. 1995. Mark–resighting analysis of a California gull population. *Journal of Applied Statistics* 22:625–639.

Reboulet, A.-M., A. Viallefont, R. Pradel, and J.-D. Lebreton. 1999. Selection of survival and recruitment models with SURGE 5.0. *Bird Study* 46:S148–S156.

Rexstad, E., and K. P. Burnham. 1991. *User's guide for interactive program CAPTURE.* Colorado Cooperative Fish and Wildlife Research Unit, Fort Collins.

Ricker, W. E. 1958. Handbook of computations for biological statistics of fish populations. *Fisheries Research Board of Canada, Bulletin 119.* 300pp.

Ricker, W. E. 1975. Computation and interpretation of biological statistics of fish populations. *Fisheries Research Board of Canada, Bulletin 191.*

Robson, D. S. 1969. Mark–recapture methods of population estimation. Pages 120–140 *in* N. L. Johnson and H. Smith, Jr., editors. *New developments in survey sampling.* Wiley, New York.

Rothery, P. 1983. Appendix: estimation of survival to breeding age in young puffins. *Ibis* 125:71–73.

Sanathanan, L. 1972. Models and estimation methods in visual scanning experiments. *Technometrics* 14:813–829.

Sandland, R. L., and P. Kirkwood. 1981. Estimation of survival in marked populations with possibly dependent sighting probabilities. *Biometrika* 68:531–541.

Schaefer, M. B. 1951. Estimation of the size of animal populations by marking experiments. *U.S. Fish and Wildlife Service Fisheries Bulletin* 69:191–203.

Schnabel, Z. E. 1938. The estimation of the total fish population of a lake. *American Mathematical Monthly* 45:348–352.

Schwarz, C. J. 2001. The Jolly-Seber model: more than just abundance. *Journal of Agricultural, Biological, and Environmental Statistics* 6:175–185.

Schwarz, C. J., M. Andrews, and M. R. Link. 1999. The Stratified-Petersen with a known number of unread tags. *Biometrics* 55:1014–1021.

Schwarz, C. J., and A. N. Arnason. 1996. A general methodology for the analysis of capture–recapture experiments in open populations. *Biometrics* 52: 860–873.

Schwarz, C. J., and A. N. Arnason. 2000. Estimation of age-specific breeding probabilities from capture–recapture data. *Biometrics* 56:59–64.

Schwarz, C. J., and J. B. Dempson. 1994. Mark–recapture estimation of a salmon smolt population. *Biometrics* 50:98–108.

Schwarz, C. J., and B. Ganter. 1995. Estimating the movement among staging areas of the barnacle goose (*Branta leucopsis*). *Journal of Applied Statistics* 22:711–724.

Schwarz, C. J., and G. A. F. Seber. 1999. Estimating animal abundance: review III. *Statistical Science* 14:427–456.

Schwarz, C. J., and W. T. Stobo. 1997. Estimating temporary migration using the robust design. *Biometrics* 53:178–194.

Schwarz, C. J., and W. T. Stobo. 1999. Estimation and effects of tag-misread rates in capture–recapture studies. *Canadian Journal of Fisheries and Aquatic Science* 56:551–559.

Schwarz, C. J., and C. G. Taylor. 1998. Use of the stratified-Petersen estimator in fisheries management: estimating the number of pink salmon (*Oncorhynchus gorbuscha*) spawners in the Fraser River. *Canadian Journal of Fisheries and Aquatic Science* 55:281–296.

Schwarz, C. J., R. E. Bailey, J. R. Irvine, and F. C. Dalziel. 1993a. Estimating salmon spawning escapement using capture–recapture methods. *Canadian Journal of Fisheries and Aquatic Sciences* 50:1181–1197.

Schwarz, C. J., J. F. Schweigert, and A. N. Arnason. 1993b. Estimating migration rates using tag–recovery data. *Biometrics* 49:177–193.

Schwarz C. J., A. N. Arnason, and C. W. Kirby. 2002. The siren song of the Schaefer estimator—no better than a pooled Petersen. Unpublished manuscript available at: http://www.stat.sfu.ca/~cschwarz/papers/2002/Schaefer/

Seber, G. A. F. 1962. The multi-sample single recapture census. *Biometrika* 49:339–350.

Seber, G. A. F. 1965. A note on the multiple recapture census. *Biometrika* 52: 249–259.

Seber, G. A. F. 1970. Estimating time-specific survival and reporting rates for adult birds from band returns. *Biometrika* 57:313–318.

Seber, G.A.F. 1982. *The estimation of animal abundance and related parameters*, 2nd ed. Chapman, London and Macmillan, New York.

Seber, G. A. F. 1986. A review of estimating animal abundance. *Biometrics* 42: 267–292.

Seber, G. A. F. 1992. A review of estimating animal abundance, II. *International Statistical Review* 60:129–166.

Seber, G. A. F. 2001. Some new directions in estimating animal population models. *Journal of Agricultural, Biological, and Environmental Statistics* 6:140–151.

Siddeek, M.S.M. 1989. The estimation of natural mortality in Irish Sea plaice, *Pleuronectes platessa* L., using tagging methods. *Journal of Fish Biology* 35:145–154.

Siddeek, M.S.M. 1991. Estimation of natural mortality of Kuwait's grooved tiger prawn *Penaeus semisulcatus* (de Haan) using tag–recapture and commercial fisheries data. *Fisheries Research (Amsterdam)* 11:109–125.

Skalski, J. R., A. Hoffman, and S. G. Smith. 1993. Testing the significance of individual- and cohort-level covariates in animal survival studies. Pages 39–49 *in* J.-D. Lebreton and P. M. North, editors. *Marked individuals in the study of bird population.* Birkhauser Verlag, Basel, Switzerland.

Smith, E. P., and G. van Belle. 1984. Nonparametric estimation of species richness. *Biometrics* 40:119–129.

Smith, P. J. 1988. Bayesian methods for multiple capture–recapture surveys. *Biometrics* 44:1177–1189.

Smith, P. J. 1991. Bayesian analyses for a multiple capture–recapture model. *Biometrika* 78:399–407.

Smith, S. G., J. R. Skalski, W. Schlechte, A. Hoffman, and V. Cassen. 1994. *SURPH.1 manual: statistical survival analysis for fish and wildlife tagging studies.* Bonneville Power Administration, Portland, OR.

Smith, T. D., J. Allen, P. J. Clapham, P. S. Hammond, S. Katona, F. Larsen, J. Lien, D. Mattila, P. J. Palsboll, J. Sigurjonsson, P. T. Stevick, and N. Oien. 1999. An ocean-basin-wide mark–recapture study of the North Atlantic Humpback Whale *(Megaptera novaeangliae). Marine Mammal Science* 15:1–32.

Spendelow, J. A., J. D. Nichols, J. E. Hines, J.-D. Lebreton, and R. Pradel. 2002. Modeling post-fledging survival and age-specific breeding probabilities in species with delayed maturity: a case study of Roseate terns at Falkner Island, Connecticut. *Journal of Applied Statistics* 29:385–405.

Stanley, T. R., and K. P. Burnham. 1998. Estimator selection for closed-population capture–recapture. *Journal of Agricultural, Biological and Environmental Statistics* 3:131–150.

Stanley, T. R., and K. P. Burnham. 1999. A goodness-of-fit test for capture–recapture model M_t under closure. *Biometrics* 55:366–375.

Stigler S. M. 1986. The history of statistics: the measurement of uncertainty before 1900. The Belknap Press of Harvard University Press, Cambridge, MA.

Stirling I., and D. Andriashek. 1992. Terrestrial maternity denning of polar bears in the eastern Beaufort Sea area. *Arctic* 45:363–366.

Stokes, S. L. 1984. The Jolly-Seber method applied to age-stratified populations. *Journal of Wildlife Management* 48:1053–1059.

Tanton, M. T. 1965. Problems of live-trapping and population estimation for the wood mouse (*Apodemus sylvaticus*). *Journal of Animal Ecology* 34:1–22.

Taylor, M. K., J. Laake, H. D. Cluff, M. Ramsay, and F. Messier. 2002. Managing the risk from hunting for the Viscount Melville Sound polar bear population. *Ursus* 13:185–202.

Viallefont, A., and J.-D. Lebreton. 1993. Estimation des tauxde survie et de migration par capture–recapture dans des populations animales stratifees spatialment. *Biometrie et analyze bes donnees spatio-temporelles, Societe Francaise de Biometrie, 12.*

Warren, W. G., and J. B. Dempson. 1995. Does temporal stratification improve the accuracy of mark–recapture estimates of smolt production: a case study based on the Conne River. *North American Journal of Fisheries Management* 15:126–131.

Wei, L. J., and D. V. Glidden. 1997. An overview of statistical methods for multiple failure time data in clinical trials. *Statistics in Medicine* 16:833–839.

White, G. C. 1983. Numerical estimation of survival rates from band recovery and biotelemetry data. *Journal of Wildlife Management* 47:716–728.

White, G. C. 2002. Discussion comments on: the use of auxiliary variables in capture–recapture modeling: an overview. *Journal of Applied Statistics* 29: 103–106.

White, G. C., and K. P. Burnham. 1999. Program MARK: survival estimation from populations of marked animals. *Bird Study* 46 (Suppl):S120–S139.

White, G. C., D. R. Anderson, K. P. Burnham, and D. L. Otis. 1982. *Capture-recapture and removal methods for sampling closed populations.* Los Alamos National Lab, LA-8787-NERP, Los Alamos, NM.

White, G. C., K. P. Burnham, and D. R. Anderson. 2001. Advanced features of program MARK. *In* R. Fields, R. J. Warren, H. Okarma, and P. R. Sievert, editors. *Wildlife, land, and people: priorities for the 21st century.* The Wildlife Society, Bethesda, MD.

Whitehead, H., and T. Arnbom. 1987. Social organizations of sperm whales off the Galapagos Islands, February–April 1985. *Canadian Journal of Zoology* 65:913–919.

Williams, B. K., J. D. Nichols, and M. J. Conroy. 2002. *Analysis and management of animal populations.* Academic Press, San Diego, CA.

Wilson, K. R., and D. R. Anderson. 1995. Continuous-time capture–recapture population estimation when capture probability varies over time. *Environmental and Ecological Statistics* 2:55–69.

Wilson, R. M., and M. F. Collins. 1993. Capture–recapture estimation with samples of size one using frequency data. *Biometrika* 79:543–553.

Wintrebert, C. 1998. Est-il possible d'etendre les tests d'adjustment des modeles de capture–recapture unistate au cas ou plusieurs stades sont observes. *Montpellier II University 51.*

Yip, P.S.F. 1991. A martingale estimating equation for a capture-recapture experiment in discrete time. *Biometrics* 47:1081–1088.

Yip, P.S.F., and A. Chao. 1996. Estimating population size from capture–recapture studies via sample coverage and estimating functions. *Communications in Statistics C—Stochastic Models* 12:17–35.

Yip, P.S.F., D.Y.T. Fong, and K. Wilson. 1993. Estimating population size by recapture sampling via estimating function. *Communications in Statistics C—Stochastic Models* 9:179–193.

Yip, P.S.F., R.M. Huggins, and D. Y. Lin. 1996. Inference for capture–recapture experiments in continuous time with variable capture rates. *Biometrika* 83: 477–483.

Yip, P.S.F., L. Xi, A. Chao, and W.-H. Hwang. 2000. Estimating the population size with a behavioral response in capture–recapture experiment. *Environmental and Ecological Statistics* 7:405–414.

Youngs, W. D. 1974. Estimation of the fraction of anglers returning tags. *Transactions of the American Fisheries Society* 103:616–618.

Youngs, W. D. 1976. An analysis of the effect of seasonal variability of harvest on the estimate of exploitation rate. *Transactions of the American Fisheries Society* 105:45–47.

Youngs, W. D., and D. S. Robson. 1975. Estimating survival rates from tag returns: model tests and sample size determination. *Journal of the Fisheries Research Board of Canada* 32:2365–2371.

Zippin, C. 1956. An evaluation of the removal method of estimating animal populations. *Biometrics* 12:163–189.

Zippin, C. 1958. The removal method of population estimation. *Journal of Wildlife Management* 22:82–90.

Contributor's Notes

RUSSELL ALPIZAR-JARA earned his Ph.D. from North Carolina State University. He is Associate Professor in the Department of Mathematics, University of Évora, Portugal, working on mathematical models for natural resources issues.

STEVEN C. AMSTRUP, as a research biologist with the US Geological Survey, has studied polar bears in Alaska for 25 years. His interests include distribution and movement patterns as well as population dynamics of wildlife and especially the ways in which information on those topics can be used to assure wise stewardship.

RICHARD J. BARKER is Associate Professor in the Department of Mathematics and Statistics at University of Otago. His research interests are in statistical ecology.

ANNE CHAO is Tsing Hua Chair Professor of Natural Science at National Tsing Hua University, Taiwan. Her main research areas include ecological statistics and related applications.

WILLIAM HEARN is a Research Scientist with CSIRO Marine Research. Bill specialises in the modeling of fisheries tagging data.

JOHN M. HOENIG is a Professor of Marine Science at the Virginia Institute of Marine Science. He has a doctorate in oceanography and a Master's degree in statistics from the University of Rhode Island. He is interested in developing quantitative methods for assessing the status of populations.

RICHARD M. HUGGINS is a Professor in the Centre for Mathematics and its Applications at the Australian National University. His research interests include the application of modern statistical techniques to the analysis of capture-recapture experiments.

TRENT L. MCDONALD is a statistician and project manager with Western EcoSystems Technology, Inc. and adjunct statistics professor at the University of Wyoming. He has co-authored a book on resource selection, and specializes in generalized linear models, ecological survey design, capture-recapture models, and computer intensive statistical methods.

BRYAN F. J. MANLY is an author of books on topics that include the statistics of natural selection, multivariate analysis, resource selection by animals, research study designs, computer-intensive statistics, and environmental statistics. He is currently a visiting professor at the University of São Paulo in Brazil. His contributions to this book were partly supported by a scholarship from CAPEs, Brazil.

JAMES D. NICHOLS is a senior scientist with the U.S. Geological Survey at Patuxent Wildlife Research Center in Laurel, Maryland. His primary research interests are the ecology and management of animal populations and methods for estimating abundance and vital rates of animal populations.

KENNETH H. POLLOCK earned his Ph.D. from Cornell University in Biometry, and is a Professor of Zoology, Biomathematics, and Statistics at North Carolina State University. He has published numerous contributions to the theory and application of capture-recapture and related sampling methods in ecology.

ERIC V. REGEHR is a PhD Candidate in the Zoology Department at the University of Wyoming. His dissertation focuses on development of better estimates and better estimators of size for wildlife populations.

CARL J. SCHWARZ is a professor at Simon Fraser University, British Columbia, Canada. His research interests are in the use of capture-recapture methods to estimate population parameters, particularly abundance. Much of his work has been applied to problems in fisheries.

Index

age, 272–73; age 0 cohort models, 105–6; age-dependent models, 46–47; age-specific breeding models, 106–7; conditional multiple-age models, 102–7; conditional single-age models, 89–102

Agresti, A., 71–72

Akaike, H., 18, 92, 98

Akaike's information criterion (AIC), 18, 47–48, 60, 72, 102, 229; Arnason-Schwarz model, 171; closed-population models, 267; conditional single-age models, 92, 98; Cormack-Jolly-Seber model, 53, 224, 236, 240, 268; covariates, 224; goodness-of-fit, 271–72; multistate models, 165; polar bears, 249; quasi, 99–100, 236, 249; reduced-parameter models, 92; tag-recovery models, 127, 139, 163; weighting of, 100

Alexander, H. M., 72

Alho, J. M., 72, 111

Allen, D. M., 130

Alpizar-Jara, Russell, 5–6, 36–57, 267–68, 301

Alt, G. L., 161

American alligator, 48–51

Amstrup, Steven C., 1–21, 301; model examples, 196–264; open-population models, 111; practical methods, 266–73

Andersen, P. K., 85

Anderson, D. R., 8, 18, 171, 254; closed-population models, 79, 84; open-population models, 92, 98–100; tag-recovery models, 131–32, 145

Arnason, A. N., 8, 198; multistate models, 166, 183, 192; open-population models, 38, 43–44, 109, 111–12

Arnason-Schwarz model, 187–88, 195; AIC, 171; Bayesian methods, 171; breeder recruitment, 175, 177; Canadian geese, 166–67, 172–76; contingency test for, 170; Cormack-Jolly-Seber model, 167–71, 175; data collection, 167–69; dead recoveries, 175; experimental protocol of, 167–69; goodness-of-fit, 169–71; individual covariates, 177;

Jolly-Seber model, 175; likelihood, 169–71; live recoveries, 175; Markovian model, 168; migration, 166–68, 170; multisample stratified open populations, 192, 194; other uses of, 172; robust design, 177

Arnbom, T., 78

Bailey, N.T.J., 22, 28, 60

bands. *See* tag-recovery systems

Banneheka, S. G., 182–84

Barker, Richard J., 7–8, 121, 142–64, 270, 301

barnacle geese, 189–93

Bayesian methods, 71, 171

Becker, N. G., 84

Bennetts, R. E., 121

Besbeas, P., 121

beta coefficients, 62, 211, 233; Bayesian methods, 71; capture probabilities 212; MARK program, 212–23; parametric approaches, 71; survival probabilities, 213

bias, 58, 224, 242, 244, 259; Bailey's binomial model, 28; catchability, 32–33; Chapman estimator, 27–28, 33, 46; classic models, 34, 41; closed-population models, 62–63, 65, 68–72, 78, 82; Jolly-Seber model, 42, 50–51, 53; model examples, multistate models, 165, 177–79, 182, 184–85; open-population models, 111, 114–15, 118; Petersen-Lincoln estimator, 27–28, 33; positive/negative population, 42–43; tag-recovery models, 134–35, 138–39, 147, 156, 160–61

biologists, 2

biometricians, 3

birth, 1, 6–7, 267–68; equations for, 40; estimation of, 40; Jolly-Seber model, 42, 47; open-population models, 36

Bjorkland, R., 116–18, 120

black bears, 21, 274

bootstrap methods, 66, 99, 239–40

Borchers, D. L., 244

Boyce, M. S., 78, 81–82, 271